ARTIFICIAL INTELLIGENCE AND MATHEMATICAL METHODS
IN PAVEMENT AND GEOMECHANICAL SYSTEMS

PROCEEDINGS OF THE INTERNATIONAL WORKSHOP ON ARTIFICIAL INTELLINGENCE AND MATHEMATICAL METHODS IN PAVEMENT AND GEOMECHANICAL SYSTEMS FLORIDA/USA/5-6 NOVEMBER 1998

Artificial Intelligence and Mathematical Methods in Pavement and Geomechanical Systems

Edited by

Nii O. Attoh-Okine
Florida International University, Miami, Florida, USA

CRC Press
Taylor & Francis Group
Boca Raton London New York

CRC Press is an imprint of the
Taylor & Francis Group, an **informa** business

A BALKEMA BOOK

Published by:
CRC Press/Balkema
P.O. Box 447, 2300 AK Leiden, The Netherlands
e-mail: Pub.NL@taylorandfrancis.com
www.crcpress.com – www.taylorandfrancis.com

© 1998 by Taylor & Francis Group, LLC
CRC Press/Balkema is an imprint of the Taylor & Francis Group, an informa business

No claim to original U.S. Government works

ISBN 13: 978-90-5809-028-7 (hbk)

**Visit the Taylor & Francis Web site at
http://www.taylorandfrancis.com**

**and the CRC Press Web site at
http://www.crcpress.com**

The texts of the various papers in this volume were set individually by typists under the supervision of each of the authors concerned.

Artificial Intelligence and Mathematical Methods in Pavement and Geomechanical Systems, Attoh-Okine (ed.)
© 1998 Balkema, Rotterdam, ISBN 90 5809 028 0

Table of contents

Artificial Intelligence and Mathematical Methods in Pavement and Geomechanical Systems, Attoh-Okine (ed.)
© *1998 Balkema, Rotterdam, ISBN 90 5809 028 0*

Preface/Forward

Over the past decade, some significant progress in the understanding of artificial Intelligence and Mathematical Methods in Pavements and Geomechanical Systems has been achieved. For example, Artificial Neural Network is becoming a standard technique in Pavement and Geomechanical Systems. The proceedings of this conference comprised of papers submitted for presentation at the International Workshop on Artificial Intelligence and Mathematical Methods in Pavement and Geomechanical Engineering Systems, held in Miami, Florida, USA from 5-6 November 1998.

Each of the papers included in the Proceedings has received at least two possible peer reviews out of the three reviewers and have been accepted for publication by the Proceedings Editor.

The Proceedings Editor thanks the authors who submitted papers, took the time to revise their manuscripts where necessary, and met the tight revision schedule.

The organizing committee wishes to thank Drs Shinya Kikuchi (Fuzzy Sets and Theory), Osama Mohammed (Genetic Algorithm) and Isaac Elishakoff (Anti-Optimization Methods Under Uncertainty) – They presented excellent tutorial sections. I would like to express my appreciation to Dr David Shen, the Chairman of the Civil and Environmental Engineering at Florida International University for providing environment conducive to the preparation of these proceedings.

The organization committee also wishes to thank Charles Nunoo, Solomon Zhrinen, Pia Hansson, Alexander Kwasi Appea, and Vivian Donis for their assistance.

Last, but not least, the Proceedings Editor dedicates these Proceedings to his mother Madam Georgina C.Quaynor (who passed away on 14 Dec. 1997, when the Editor was planning this workshop).

Nii O.Attoh-Okine, Ph.D., P.E.
Editor

Artificial Intelligence and Mathematical Methods in Pavement and Geomechanical Systems, Attoh-Okine (ed.)
© *1998 Balkema, Rotterdam, ISBN 90 5809 028 0*

Uses of artificial neural networks in the mechanistic-empirical design of flexible pavements

R.W. Meier
Department of Civil Engineering, The University of Memphis, Tenn., USA

E. Tutumluer
Department of Civil Engineering, University of Illinois at Urbana-Champaign, Ill., USA

ABSTRACT: The AASHTO Joint Task Force on Pavements recently initiated an effort to develop, by the year 2002, an improved design guide based on mechanistic-empirical design principles. That effort is the impetus for this examination of existing and potential uses of artificial neural networks in mechanistic-empirical pavement design. Many of the tasks at which ANNs excel—optimization, function approximation, pattern recognition, and forecasting are just a few—are present in mechanistic-empirical pavement design. This paper discusses the specific areas of mechanistic-empirical pavement design to which ANNs may be applied and the types of ANNs most suited to those tasks. This paper shows that ANNs can be a useful complement to more-traditional numerical and statistical methods and that their use in pavement design should continue to be investigated.

1 INTRODUCTION

Mechanistic-empirical (M-E) principles have been used for more than 40 years in the design of rigid pavements, but have not yet seen widespread use in designing flexible pavements. Traditional methods of flexible pavement design, such as those presented in the widely-used AASHTO *Guide for Design of Pavement Structures* (1993), rely on empirical correlations between pavement performance and structural design parameters. Those correlations are difficult to apply to loading conditions, material properties, and construction methods that differ from those underlying the original data. This has induced pavement designers to search for methods that are more universally applicable, such as those based on theoretical mechanics.

In M-E design, pavement response variables, such as the stresses, strains and deflections produced by wheel loads, are first calculated using the principles of theoretical mechanics. Those response variables are then related to measures of pavement performance or distress using transfer functions developed from laboratory tests and field performance data. The transfer functions help to account for factors not precisely modeled by theory alone while the theoretical analysis makes the procedure broadly applicable. The transition to M-E design is being facilitated by the availability of high-speed microcomputers to perform the theoretical analyses and sophisticated laboratory and field tests to provide the necessary constitutive data.

Some state highway agencies, such as those in Kentucky (Southgate & Deen 1987) and Illinois (Illinois Department of Transportation, 1995.), have already established M-E design procedures for new pavements. M-E design methods for flexible pavements have also been developed by industry groups such as the Asphalt Institute (1982) and Shell (Claussen *et al.* 1977) and federal agencies such as the FAA (U.S. Department of Transportation, 1995). The AASHTO Joint Task Force on Pavements (JTFOP), which is responsible for the development and implementation of pavement design technologies, recently initiated an effort to develop an improved design guide by the year 2002. The 2002 AASHTO Design Guide will be developed under the oversight of NCHRP Project Panel C1-37 and, to the largest extent possible, will be based on sound mechanistic principles for the design of new and rehabilitated pavement structures.

Over the past two decades, there has also been increasing interest in artificial neural networks (ANNs). ANNs are powerful and versatile computational tools that can be used to organize and correlate information in ways that have proved useful for solving certain types of problems too complex, too poorly understood, or too resource-intensive to tackle using more-traditional computational methods. ANNs have been successfully used for tasks involving pattern recognition, function approximation, optimization, forecasting, data retrieval, and automatic control, to name just a few. Because many of these tasks are inherent in M-E design, it seems natural to investigate the application of ANNs to M-E design.

The earliest applications of ANNs in transportation engineering concentrated on areas such as planning, traffic control and operations, construction and maintenance, and facilities management (Faghri, *et al.* 1997; & Dougherty, 1995). ANNs have also been successfully applied in pavement-related areas such as foundation engineering and material modeling (Toll, 1996). The last few years have seen considerable interest in using ANNs for pavement analysis and design. The research reviewed here shows that ANNs are applicable to almost all aspects of M-E design. This doesn't mean that ANNs are the *only* tools that can be used in M-E design; only that they should not be overlooked in the development of M-E design procedures.

2 MECHANISTIC-EMPIRICAL DESIGN

The first mechanistic design curves for flexible pavements, based on elastic layered theory, were developed in the early 1960s (Dormon & Metcalf 1965). Lacking the sophisticated computational resources we enjoy today, each curve had to be laboriously calculated by hand, so they could only be developed for a limited range of idealized pavement systems. Since then, the computational resources with which to calculate wheel-load-induced pavement responses in multi-layered pavement systems have increased tremendously. This has made it much more feasible to employ mechanistic analysis procedures in pavement design.

NCHRP Project 1-26 (National Research Council 1990, 1992.) was the first sponsored research project to be initiated with the objective of developing mechanistic pavement analysis and design procedures suitable for use in future versions of the AASHTO guide. After identifying and evaluating the available technologies, NCHRP 1-26 researchers proposed working versions of M-E design processes and procedures that relate pavement responses to the development of specific types of pavement distress.

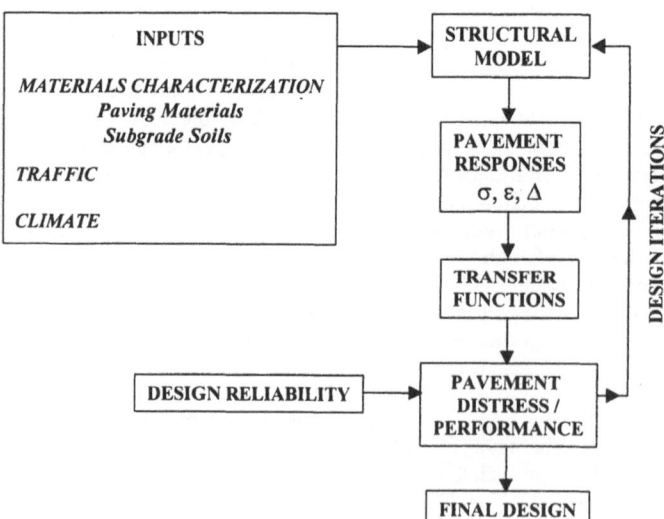

Figure 1. Components of mechanistic-empirical pavement design (after NCHRP 1-26 Phase I, 1990)

The general components of an M-E design procedure are illustrated in Figure 1. The two major components are 1) a pavement structural model to calculate pavement responses, and 2) transfer functions to translate those responses into measures of pavement performance. The design process entails iteratively adjusting the pavement structure until the desired level of performance and reliability are achieved.

To calculate pavement structural response, it is necessary to first obtain information about the pavement materials, the environment (climate), and the traffic loads. The pavement materials are characterized by strength and stiffness properties that can be obtained directly from laboratory tests or backcalculated from nondestructive tests conducted in situ. Climatic effects include temperature, moisture, and drainage conditions. The loading conditions are defined by the magnitude, frequency, and location of the wheel loads applied to the pavement. Because all of these inputs to the structural model change over time, so does the pavement's response. M-E design procedures must account for these changes throughout the service life of the pavement.

Recommendations developed in NCHRP 1-26 Phase I indicated that the elastic layer programs (ELPs), such as BISAR (DeJong *et al.* 1973), JULEA (Federal Aviation Administration, 1995), and WESLEA (Van Cauwelaert *et al.*, 1986), and finite element structural models, such as ILLI-PAVE (Raad & Figueroa 1992), GT-PAVE (Tutumluer 1995), ILLI-SLAB (Tabatabaie & Barenberg 1980), are adequate to support the development of mechanistic-empirical pavement thickness design procedures. In the ELPs, which are computationally much simpler than the finite-element models (FEMs), pavement materials are assumed to be linearly elastic, isotropic, and homogeneous within well-defined horizontal layers. The joints within a Portland cement concrete pavement naturally conflict with those assumptions. So, too, do the stress- and rate-dependent mechanical properties of many pavement materials. The FEMs, on the other hand, easily accommodate irregular geometries and stress-dependent material properties, but require vastly greater amounts of computer resources and a more-detailed characterization of the pavement materials.

The transfer functions (also called distress models) relate the computed pavement responses to pavement performance as measured by the type, severity, and extent of distress (e.g., rutting, cracking, and ride roughness). The most commonly used transfer functions relate pavement life to flexural stress or strain levels and to surface or subgrade deflections. NCHRP 1-26 researchers concluded that the transfer functions are the weak link in the M-E design approach. Predicted and observed pavement distress and performance often do not compare favorably. Extensive field calibration and verification will be required to establish reliable distress prediction models. This will have to be done locally (on a state-by-state basis) to account for differences in material properties and climate that are not explicitly modeled.

3 ARTIFICIAL NEURAL NETWORKS

The term "artificial neural networks" encompasses a wide array of computational tools loosely patterned after biological processes. Physically, all ANNs are interconnected assemblages of mathematically simple computational elements. These computational elements contain a very limited amount of local memory and perform rudimentary mathematical operations on data passing through them. The computational power of ANNs comes from parallelism—input data is concurrently operated upon by multiple computational elements.

Functionally, all ANNs are "vector mappers" (Wasserman, 1993) that accept a feature vector from one data space and produce from it an associated feature vector in another data space. Hopfield (1982) referred to this as "emergent computation" because the input vectors disappear into the network, becoming unidentifiable once inside, then emerge as output. Inside the network, data passes between computational elements along weighted connections. Because the data that emerges from the network changes as the connection weights change, ANNs can "learn" to produce a desired output by adjusting the signs and magnitudes of their weights. The appropriate adjustments are determined by the computational elements themselves using learning rules that seek to minimize some type of cost or energy function. Each computational element simply works to improve its own performance. In the process, the performance of the network as a whole is optimized for the task at hand. This "parallel

3

distributed processing" (Rumelhart *et al.*, 1986) gives ANNs the ability to learn very complex mappings without having to specify a priori functions and rules, as is required by conventional computational methods. The user needs only select the correct type of network and the most appropriate data representation (i.e., feature vectors) for the problem being solved.

There are nearly as many different types of ANN as there are researchers working in the field. They differ in the arrangement and degree of connectivity of their computational elements, the types of calculations performed within each computational element, the degree of supervision they receive during training, the determinism of the learning process, and the overall learning theory under which they operate (Mehra & Wah, 1992). Despite that, certain types of ANN appear over and over again either because they are broadly applicable to a wide variety of problems or ideally suited for a narrow range of problems. Several of these are briefly discussed next in roughly the order in which they were introduced.

Hopfield nets (Hopfield, 1984; Hopfield & Tank, 1986) store a set of patterns (feature vectors) in such a way that the network, when presented with a new pattern, responds with the stored pattern that most closely resembles the new pattern. The Hopfield net actually implements an energy function in which each stored pattern is a local minimum. Any new pattern introduced to the network will follow the surface of that energy function to the nearest local minimum—the stored pattern that most closely matches it.

Hopfield nets can be used for pattern recognition (selecting one pattern from a set of possible matches), pattern completion (providing a complete pattern from incomplete data), classification (identifying a pattern as belonging to a specific group), and as a content-addressable memory (retrieving complete records given partial information from those records). If properly designed, Hopfield nets can also be used for optimization. If the optimization problem can be written as a Hopfield energy function, the network can find a near-optimal solution to the problem given any starting point. Hopfield nets have been very successfully used for finding near-optimal solutions to combinatorial optimization problems such as the traveling salesman problem (Hopfield & Tank, 1985).

Adaptive Resonance Theory (ART) networks (Carpenter & Grossberg 1987, 1988) store a set of patterns in such a way that the network, when presented with a new pattern, either matches it to a previously stored pattern or, if it is not sufficiently similar to any of the existing patterns, stores it as a new pattern prototype to which future patterns can be matched. This allows ART networks to evolve over time as they are presented with new data. This process is called "unsupervised learning" because the network adapts to its information environment without intervention.

ART networks, like Hopfield networks, can be used for pattern recognition, completion, and classification and as a content-addressable memory. They can also be used for knowledge processing (i.e., organizing existing knowledge into groups and identifying new knowledge). In this last capacity, they could be used to detect anomalies in data because the creation of a new pattern prototype indicates an anomalous feature vector.

Kohonen maps (Kohonen 1982), also called feature maps, self-organize to produce consistent outputs for similar inputs. Specifically, Kohonen maps take data (feature vectors) from one data space and project them into a lower-ordered data space (usually a line or plane) in such a way that similar feature vectors project onto points in close proximity to one another. This is called topology preservation.

Kohonen maps can be used for pattern recognition and classification and for data compression (data is mapped into a space with fewer dimensions while preserving as much content as possible). To illustrate this, think about the colors on a computer screen. By combining red, green, and blue in varying amounts, millions of colors can be created. Each color is actually a feature vector in 3-dimensional (RGB) space. A Kohonen map can take those 3-dimensional color inputs and project them onto a 2-dimensional plane with a finite number of "pixels" so that all the yellows cluster together, all the purples cluster together, and so on. The 3-tuple describing each color input has been replaced by the (x,y) location of the pixel that most closely approximates the color.

Backpropagation networks are the workhorses of ANNs. They are very powerful and versatile networks that can be "taught" a mapping from one data space to another using examples of the mapping to be learned. The term "backpropagation network" actually refers to a multi-layered, feed-forward neural network (Hecht-Nielsen, R., 1990) trained using an error-backpropagation algorithm (Werbos, 1974; Parker, 1985; Rumelhart *et al.* 1986). As with many ANNs, the connection weights are initially selected at random. Inputs from the mapping examples are propagated forward through each layer of the network to emerge as outputs. The errors between those outputs and the correct answers are then propagated backwards through the network and the connection weights are individually adjusted so as to reduce the error. After many examples have been propagated through the network many times, the mapping function is "learned" to within some error tolerance. This is called supervised learning because the network has to be shown the correct answers in order for it to learn. Backpropagation networks excel at data modeling and classification. They have also been successfully used for image compression (they're taught to map the inputs back onto themselves), forecasting, and pattern recognition (Hertz *et al.* 1991).

Counterpropagation networks (Hecht-Nielsen 1987-1988) are hybrid networks that combine supervised and unsupervised learning to create a self-organizing look-up table that can be used for function approximation and classification. As input feature vectors from a training set are presented to the network, unsupervised learning is used to create a topology-preserving (Kohonen) map of the input data while, at the same time, supervised learning is used to associate an appropriate output feature vector with each point on the map. The output at each point is just the average output for all of the feature vectors that map to that point.

Once the network has been trained, each new feature vector presented to the network will trigger a response that is the average for those feature vectors closest to it in the input data space. This is the function of a look-up table. The advantage this network has over conventional look-up tables is that the Kohonen map provides for a statistically optimal coverage of the input space even if the mathematical form of the underlying function is completely unknown. Counterpropagation networks train much faster than backpropagation networks but are not as versatile and are comparatively slower at producing outputs.

Radial Basis Function Networks (Moody & Darken 1988,1989; Poggio & Girosi 1990) are also hybrids that combine unsupervised and supervised learning to perform function approximation. The concept is really rather simple: summing a series of overlapping Gaussian functions can approximate any continuous function. In two dimensions, Gaussian curves are familiar to many as the normal distribution from statistics. In three dimensions, they appear as "bumps" with radial symmetry. In higher dimensions, they are difficult to visualize but the concept is equally valid.

The radial basis function network has a mapping layer in which each neuron represents one Gaussian bump. As with the counterpropagation network, unsupervised learning is used to determine how best to partition the data space given a limited number of neurons. Each neuron is assigned to a cluster of input vectors and affects a Gaussian bump located at the center of the cluster. Once the data space has been appropriately partitioned, supervised learning is used to adjust the heights of the bumps so as to produce the best approximation of the function. When a new input vector is presented to the trained network, it responds with an output that is really just the sum of the outputs from every Gaussian bump in the network weighted according to the distance from the input vector to the centers of the bumps.

Radial basis function networks also train much faster than backpropagation networks but are not as versatile and are comparatively slower in use because each output requires that dozens (or even hundreds) of Gaussian functions be evaluated.

Generalized Regression Neural Networks (GRNNs) (Specht, 1991) are closely related to radial basis function networks. In a GRNN, each neuron in the mapping layer represents a Gaussian bump that coincides exactly with one of the inputs from the training set. Since there is exactly one neuron for each training example, the weights are simply set by hand using the input and output feature vectors for each example. The training time is therefore zeroed—the weights are initialized to the coordinates of the feature vectors in the training set. Unfortunately, because the training examples do not optimally cover the input space, many of the neurons are wasted, so more neurons are needed to achieve the same error level as a radial basis function network. This makes them even slower than radial basis function networks at producing an output.

5

4   CURRENT AND POTENTIAL USES FOR ANNS

As might be expected from the wide variety of network types above, there are many different areas in which ANNs can be used to support M-E design initiatives. This section will detail applicable work that has already been done and identify some of the tasks for which ANNs are particularly well suited and should be investigated. We'll begin with the transfer functions because they are currently regarded as the weak link in the M-E design procedure.

4.1   Transfer Functions

There is great potential for using ANNs in the development of predictive distress models. They are exactly the type of complex, multivariate problem that has been successfully solved using ANNs. Any of the function approximation neural networks (backpropagation, radial basis function, counterpropagation, and others) could potentially be used to develop correlations between structural response variables and measures of pavement distress. The key to success is not so much the type of network as the volume and quality of training data.

Some state highway agencies, such as those in Illinois and the North Carolina, have begun monitoring selected in-service pavements for performance. They are keeping records of pavement materials and cross-sections, applied traffic loads, and climatic conditions. Similar data will eventually be generated, in even greater volumes, from the Long Term Pavement Performance (LTPP) project of the Strategic Highway Research Program. The structural response of the pavements to the recorded loads can be calculated using mechanistic analysis programs. The observed pavement performance can then be correlated with the calculated structural response using ANNs. Because ANNs excel at mapping in higher-order spaces, such models can go beyond the existing univariate relationships (such as those based on asphalt flexural strain or subgrade vertical compressive strain). ANNs could be used to examine several variables at once and the interrelationships between them. ANNs could also be used to develop models for distress phenomena such as thermal cracking, block cracking, and rutting in AC pavements, and faulting and D-cracking in concrete pavements.

4.2   Pavement Distresses

Before any transfer functions can be developed, quantitative measures of pavement distress and performance must be established and methods developed to measure their values over time. This is an area in which neural networks have already shown promises. Kaseko *et al.* (1994) successfully used both a backpropagation network and a proprietary ANN (Lo & Bavarian, 1991) to automatically detect and classify various types of surface cracks in video images of AC pavements. Chou *et al.* (1994, 1995) used backpropagation networks to classify surface cracks, which had already been extracted from the video images by other means. Wang (1995) proposes using a neural net computer chip—the Intel Ni1000, which implements a radial basis function network—to automatically detect, classify, and quantify different types of pavement distress at highway speeds.

4.3   Shift Factors

Transfer functions, whether they are developed using ANNs or conventional modeling techniques, will have to be calibrated to local conditions using shift factors. Shift factors are used to adjust the predicted distress development to more realistically reflect field-observed pavement distress and performance. These shift factors not only vary from state to state, but will have to be periodically updated for temporal changes in climate, materials, construction specifications, and traffic. Radial basis function networks would be particularly well-suited to this task because they can be incrementally retrained. This is an important point. Some ANNs, such as backpropagation networks, must be completely retrained if additional data becomes available. Others, such as radial basis function and counterpropagation networks, can evolve over time to accommodate both new data and changed data. The choice of a network type should anticipate future enhancements as well as present needs.

4.4   Pavement Performance Tests

Another application for which ANNs are well suited is the reduction and processing of the huge databases developed from pavement performance and accelerated pavement-testing projects. There are currently several such projects underway, such as the nationwide LTPP program, the Westrack project in Nevada, and the Minnesota DOT MnRoad Project. The comprehensive databases developed from these studies will contribute significantly to the development of improved pavement transfer functions. Neural networks can be used to facilitate the reduction, analysis, and correlation of this data.

Banan and Hjelmstad (1996) illustrated the potential of ANNs in this area by re-examining the AASHO Road test data using a proprietary ANN which, like a radial basis function network, subdivides the input space and learns an average response for each subdivision Banan and Hjelmstad (1995). By fitting the data locally, they were able to obtain much better data correlations rather than were obtained using regression, which globally fits the data to a single mathematical function.

4.5   Pavement Structural Models

Despite exponential advances in computational speed, pavement structural models still expend considerable amounts of computing time. Even the slowest ANNs are two or three orders of magnitude faster than ELPs and several more orders of magnitude faster than FEMs. There are several ways in which function approximation ANNs can be used to speed the structural analysis task.

Meier *et al.* (1997) trained backpropagation networks as surrogates for WESLEA in a computer program for backcalculating pavement layer moduli and realized a fifty-fold increase in processing speed. Backpropagation networks were chosen because they are faster than the other function approximation networks, albeit much harder to train. This is an important consideration when choosing what network to use.

ANNs will never completely replace the versatility of an FEM or ELP, but can be suitable surrogates as long as the mapping problem is reasonably constrained. It would be unrealistic to expect an ANN to compute stresses, strains, and deflections anywhere in a pavement under any loading conditions. On the other hand, it would be relatively easy to train an ANN to compute, for example, the maximum tensile strain at the bottom of the asphalt-bound layers of a flexible pavement due to a dual wheel load. Ceylan *et al.* (1998) illustrated this capability by training an ANN to compute lateral and longitudinal tensile stresses at the bottom of jointed concrete airfield pavements as a function of load location, slab thickness, subgrade support, and the load transfer efficiencies of the joints. The training set was developed for a prescribed dual wheel load using the ILLI-SLAB finite element program.

Haussmann *et al.* (1997) propose a different approach to speeding the structural analysis task. As previously mentioned, ELPs are several orders of magnitude faster than FEMs but are inappropriate for analyzing jointed concrete pavements because the underlying assumptions of layer continuity are violated. They've proposed using ANNs to correct ELP results for the effects of finite slab size. In effect, the ANNs would be taught the *difference* between the ELP and FEM results as a function of the geometry of the slab and the position of the wheel loads relative to the slab edges. This same approach could be extended to FEM analysis of flexible pavements that includes discontinuities such as cracks, joints, and shoulders.

4.6   Material Characterization

Pavement structural models require that the constitutive behavior of the pavement materials be determined. Material characterization actually covers a wide range of topics from the measurement of elastic material properties in lab or field to mathematical modeling of nonlinear elastoplastic or viscoelastic constitutive behavior. ANNs have already been applied in both of those areas.

Meier and Rix (1994, 1995) trained backpropagation networks to backcalculate AC pavement layer moduli from deflection basins obtained using the falling weight deflectometer (FWD). They achieved their goal of increased backcalculation speed by creating a neural network that operates 4500 times faster than the conventional algorithmic program being used at the time. Khazanovich and Roesler (1997) used a proprietary neural network to perform the same task for data obtained from composite pavements. Ioannides et al. (1996) used a backpropagation neural network to determine the load transfer efficiency of rigid pavement joints from FWD data. Rolling wheel deflectometers (RWD) are currently under development which will make it possible to induce and measure deflection basins at realistic traffic speeds. Neural networks may be the only method fast enough to analyze the enormous volumes of data that will be generated in a reasonable amount of time.

Williams and Gucunski (1995) and Gucunski and Krstic (1996) investigated backpropagation and GRNN networks for simultaneously backcalculating both layer moduli and layer thicknesses from dispersion curves generated by Spectral Analysis of Surface Waves (SASW) tests. Meier and Rix (1993) had previously shown that neural networks could be used to backcalculate the moduli and thicknesses of soil layers from the results of SASW tests. Though currently more of a research tool than a production tool, SASW is an attractive alternative to FWD and could be significant in the future.

Compared to backcalculation, there has been relatively little work done on constitutive modeling of pavement materials using ANNs. Tutumluer and Meier (1996) attempted to develop a universal stress-dependent model for the resilient modulus of unbound granular base materials using neural networks. Their paper clearly illustrates the need for a large volume of carefully selected training data and a statistically independent testing set with which to validate the trained network. It is not sufficient to simply pick a handful of points at random.

Tutumluer and Seyhan (1998) recently trained a backpropagation ANN to predict the anisotropic stiffness properties of granular materials from standard repeated load triaxial tests (i.e., tests lacking lateral deformation measurements). The ANN was trained using laboratory test data. The inputs consisted of confining pressure, applied deviator stress, measured vertical deformation, compacted dry density, and percentage of crushed particles. The outputs were the horizontal and shear moduli.

The majority of ANN-based constitutive models in the literature are for soils rather than paving materials. Ellis et al. (1995) developed an ANN-based constitutive model for sands based on grain size distribution and stress history. Penumadu et al. (1994) developed an ANN-based constitutive model that captured the rate-dependent behavior of clay soils. Zhu and Zaman (1997) recently trained a recurrent neural network (a variant of the backpropagation network often used for time-series analysis) to accurately predict the axial and volumetric stress-strain behavior of sand during loading, unloading, and reloading. In a similar application, Basheer and Najjar (1998) trained recurrent ANN models for characterizing the cyclic behavior of fine-grained soils.

There is still considerable work to be done in this area. ANNs could be used to model the temperature- and rate-dependent behavior of asphaltic concrete, the anisotropy of granular base materials, or the fatigue behavior of asphaltic and portland cement concrete. Models such as these are especially needed because M-E design procedures will be based on the structural analysis of pavements throughout their design life. This requires that changes in the behavior of the pavement materials over time, due to such things as seasonal variations in temperature and moisture and to fatigue, be modeled explicitly.

4.7   Traffic Loads

The final area of structural analysis in which ANNs could be used is the modeling of the traffic loads applied to the pavement. Unlike current empirical design procedures based on equivalent axle loads or equivalent aircraft, M-E design procedures make it possible to explicitly model the landing gear geometries and wheel loads of each individual vehicle. This requires the user to develop a realistic vehicle mix and project it over the design life of the vehicle. In that capacity, ANNs developed for time series analysis (such as recurrent neural networks) can play a major role.

Another area in which ANNs have already been used is roadway classification. Most state highway agencies maintain permanent traffic recorder stations at strategic locations in order to develop a database of traffic patterns for different road types. This database captures the seasonal variations in the monthly average daily traffic (MADT) at each recorder location. In principle, if you can match a road segment where there is no recorder to one stored in the database, you can forecast its average annual daily traffic (AADT) from a short-term traffic count by applying the seasonal variations stored in the database. In practice, the database is condensed into a handful of road types exhibiting similar traffic patterns and road attributes; this makes it easier to find a match in the database. The task of condensing the database is a classic pattern classification application at which ANNs excel.

Faghri and Hua (1995) used an ART network to group road segments according to their MADT patterns. Their ANN produces groups with far less data scatter than groups created using conventional methods of cluster analysis and regression analysis. Lingras (1995) used Kohonen maps for a similar purpose and found them to be a suitable alternative to hierarchical classification methods currently being used. The Kohonen maps were able to classify both complete and incomplete traffic patterns and to evolve over time as traffic patterns changed.

5   SUMMARY

Artificial neural networks (ANNs) are valuable computational tools for solving problems too complex, poorly understood, or resource-intensive to tackle using more-traditional methods. Many of the tasks at which they excel, such as pattern recognition, function approximation, optimization, and forecasting, are inherent in mechanistic-empirical pavement design. This paper has identified specific areas in which ANNs might be successfully applied and discussed the types of ANNs most suited to those tasks. It has also provided many examples of successful ANN applications in the area of flexible pavement design. ANNs have been shown to be a very useful complement to more-traditional computational methods that clearly should be investigated further.

REFERENCES

AASHTO Guide for Design of Pavement Structures, 1993, Washington DC.

Airport Pavement Design for the Boeing 777 Airplane. 1995 Federal Aviation Administration, Advisory Circular (AC) No: 150/5320-16. U.S. Department of Transportation, Washington D.C.

Banan, M. R. & Hjelmstad, K. D., 1995. A Monte Carlo Strategy for Data-Based Mathematical Modeling, *Journal of Mathematical and Computer Modeling*, Vol. 22, No. 8, pp. 73-90.

Banan, M. R. & Hjelmstad, K. D.. September/October 1996. Neural Networks and the AASHO Road Test, *Journal of Transportation Engineering, ASCE*, Vol. 122, No. 5, , pp. 358-366.

Basheer, I.A. & Najjar, Y.M., 1998. Modeling Cyclic Constitutive Behavior by Neural Networks: Theoretical and Real Data, Proceedings of the 12th Engineering Mechanics Conference (H. Murakami & J.E. Luco, Ed.), La Jolla, California, May 17-20, pp. 952-955.

Carpenter, G. A. & Grossberg, S., 1987. A Massively Parallel Architecture for a Self-Organizing Neural Pattern Recognition Machine, Computer Vision, Graphics, and Image Processing, Vol. 37, pp. 54-115.

Carpenter, G. A. & Grossberg, S., 1987. ART2: Self-Organization of Stable Category Recognition Codes for Analog Input Patterns, *Applied Optics*, Vol. 26, pp. 4919-4930.

Carpenter, G. A. & Grossberg, S., March 1988. The ART of Adaptive Pattern Recognition by a Self-Organizing Neural Network, *Computer*, pp. 77-88.

Ceylan, H., Tutumluer, E., & Barenberg, E.J., June 14-17, 1998. Artificial Neural Networks As Design Tools in Concrete Airfield Pavement Design, Proceedings of the International Air Transportation Conference, Austin, Texas.

Chou, J., O'Neill, W. A., & Cheng, H. D., 1994. Pavement Distress Classification Using Neural Networks Proceedings, IEEE International Conference on Systems, Man, and Cybernetics, Vol. 1, pp. 397-401.

Chou, J., O'Neill, W. A., & Cheng, H. D., 1995. Pavement Distress Evaluation Using Fuzzy Logic and Moment Invariants, *Transportation Research Record 1505*, Transportation Research Board, , pp. 39-46.

Claussen, A.I.M., Edwards, J.M., Sommer, P., & Uge, P., 1977. Asphalt Pavement Design -The Shell Method. In Proceedings of the 4th International Conference on the Structural Design of Asphalt Pavements, Vol. 1, , pp. 39-74.

DeJong, D. L., Peutz, M. G. F., & Korswagen, A. R., 1973. Computer Program BISAR: Layered Systems Under Normal and Tangential Loads, External Report AMSR.0006.73, Koninklijkel/Shell-Laboratorium, Amsterdam, the Netherlands.

Design Manual 1995. *Conventional Flexible Pavement Design for Local Agencies*. Bureau of Local Roads and Streets Illinois Department of Transportation, Springfield.

Dormon, G.M. & Metcalf, C.T., 1965. Design Curves for Flexible Pavements Based on Layered Systems Theory. *Highway Research Record 71*, Highway Research Board, pp. 69-84.

Dougherty, M., 1995. A Review of Neural Networks Applied to Transport. Transportation Research –C, Vol. 3, No. 4, pp. 247-260.

Ellis, G. W., Yao, C., Zhao, R. & Penumadu, D., 1995. Stress-Strain Modeling of Sands Using Artificial Neural Networks, Journal of Geotechnical Engineering, ASCE, Vol. 121, No. 5, pp. 429-435.

FAA Advisory Circular (AC) No: 150/5320-16, 1995. Airport Pavement Design for the Boeing 777 Airplane, Federal Aviation Administration, Washington, D.C.

Faghir, A. & Hua, J., July 1995. Roadway Seasonal Classification Using Neural Networks, *Journal of Computing in Civil Engineering, ASCE,* Vol. 9, No. 3, pp. 209-215.

Faghri, A., Martinelli, D., & Demetsky, M.J.,1997. Chapter 7: Neural Network Applications in Transportation Engineering. In Artificial Neural Networks for Civil Engineers – Fundamentals and Applications, ed. Nabil Kartam, Ian Flood, & James H. Garrett, Jr. Expert Systems and Artificial Intelligence Committee, ASCE.

Full-Depth Bituminous Concrete Pavement Design for Local Agencies., 1995. Design Manual, Bureau of Local Roads and Streets, Illinois Department of Transportation, Springfield.

Gucunski, N. & Krstic, V., 1996. Backcalculation of Pavement Profiles from Spectral-Analysis-of-Surface Waves Test by Neural Networks Using Individual Receiver Spacing Approach, *Transportation Research Record 1526,* TRB, National Research Council, Washington, D. C, pp. 6-13.

Haussmann, L. D., Tutumluer, E., & Barenberg, E. J., January 1997. Neural Network Algorithms for the Correction of Concrete Slab Stresses from Linear Elastic Layered Programs, Preprint No. 970540, 76th Meeting of the Transportation Research Board, Washington, D. C.

Hecht-Nielsen, R., 1987. Counterpropagation Networks, *Applied Optics*, Vol. 26, , pp. 4979-4984.

Hecht-Nielsen, R., 1988. Applications of Counterpropagation Networks, Neural Networks, Vol. 1, , pp. 131-139.

Hecht-Nielsen, R. 1990. Neurocomputing, Addison-Wesley, New York,.

Hertz, J., Krogh, A., & Palmer, R. G., 1991. Introduction to the Theory of Neural Computation, Addison-Wesley, New York, pp. 130-141.

Hopfield, J. J., Neural April 1982. Networks and Physical Systems with Emergent Collective Computational Abilities, Proceedings, National Academy of Sciences, Vol. 79, National Academy of Sciences, Washington, D. C., pp. 2554-2558.

Hopfield, J. J., , May 1984. Neurons with Graded Response Have Collective Computational Properties Like Those of Two-State Neurons, Proceedings, National Academy of Sciences, Vol. 81, National Academy of Sciences, Washington, D. C., pp. 3088-3092.

Hopfield, J. J. & Tank, D. W., 1985. "Neural" Computation of Decisions in Optimization Problems, *Biological Cybernetics*, Vol. 52, pp. 141-152.

Hopfield, J. J. & Tank, D. W., August 1986. Computing with Neural Circuits: A Model, *Science*, Vol. 233, pp. 625-633.

Ioannides, A. M., Alexander, D. R., Hammons, M. I., & Davis, C. M., ., 1996. Application of Artificial Neural Networks to Concrete Pavement Joint Evaluation, *Transportation Research Record* No. 1540, TRB, National Research Council, Washington, D. C, pp. 56-64.

Kaseko, M. S., Lo, Z-P, & Ritchie, S G., July/August 1994. Comparison of Traditional and Neural Classifiers for Pavement Crack Detection, *Journal of Transportation Engineering, ASCE*, Vol. 120, No. 4, pp. 552-569.

Khazanovich, L. & Roesler, J., January 1997. DIPLOBACK: A Neural-Networks-Based Backcalculation Program for Composite Pavements, Preprint No. 970752, 76th Meeting of the Transportation Research Board, Washington, D. C.

Kohonen, T., , 1982. Self-Organized Formation of Topologically Correct Feature Maps, *Biological Cybernetics*, Vol. 43, pp. 59-69.

Lingras, P., July/August 1995. Classifying Highways: Hierarchical Grouping Versus Kohonen Neural

Networks, *Journal of Transportation Engineering, ASCE*, Vol. 121, No. 4, , pp. 364-368.

Lo, Z. P. & Bavarian, B., , 1991. A Neural Piecewise Linear Classifier for Pattern Classification, Proceedings, IEEE International Joint Neural Network Conference, Vol. 1, pp. 264-268.

Mehra, P. & Wah, B. W., 1992. Artificial Neural Networks: Concepts and Theory, IEEE Computer Society Press, Los Alamitos, CA, pp. 1-8.

Meier, R. W. & Rix, G. J., December 1993. An Initital Study of Surface Wave Inversion Using Artificial Neural Networks, *Geotechnical Testing Journal*, ASTM, Vol. 16, No. 4.

Meier, R. W. & Rix, G. J., 1994. Backcalculation of Flexible Pavement Moduli Using Artificial Neural Networks, *Transportation Research Record No. 1448*, TRB, National Research Council, Washington, D. C., pp. 75-82.

Meier, R. W. & Rix, G. J., 1995. Backcalculation of Flexible Pavement Moduli from Dynamic Deflection Basins Using Artificial Neural Networks, *Transportation Research Record No. 1473*, TRB, National Research Council, Washington, D. C., pp. 72-81.

Meier, R. W., Alexander, D. R. & Freeman, R., January 1997. A Forward Approach to Backcalculation using Artificial Neural Networks, Preprint No. 970235, 76th Meeting of the Transportation Research Board, Washington, D. C.

Moody, J. and Darken, C., 1988. Learning with Localized Receptive Fields, Proceedings of the 1988 Connectionist Models Summer School, Morgan Kaufmann, San Mateo, , pp. 133-143.

Moody, J. & Darken, C., 1989. Fast Learning in Networks of Locally-Tuned Processing Units, *Neural Computation*, Vol. 1, , pp. 281-294.

NCHRP Project 1-26, Final Report, Phase 1 1990. *Calibrated Mechanistic Structural Analysis Procedure for Pavements*, Transportation Research Board, National Research Council, Washington D.C.

NCHRP Project 1-26, Final Report, Phase 2 1992. *Calibrated Mechanistic Structural Analysis Procedure for Pavements*. Transportation Research Board, National Research Council, Washington D.C.

Parker, D. B., 1985. Learning Logic, Technical Report TR-47, Center for Computational Research in Economics and Management Science, Massachusetts Institute of Technology, Cambridge, MA,.

Penumadu, D., Jin-Nan, L., Chameau, J-L., & Arumugam, S., , 1994. Rate Dependent Behavior of Clays Using Neural Networks, Proceedings of the 13th Conference of the International Society for Soil Mechanics and Foundation Engineering, Oxford and IBH Publishing Co., Vol. 4, pp. 1445-1448.

Poggio, T. & Girosi, F., 1990.Regularization Algorithms for Learning that are Equivalent to Multilayer Networks, *Science*, Vol. 247, pp. 978-982.

Raad, L. & Figueroa, J.L., Load January 1980. Response of Transportation Support Systems. *Transportation Engineering Journal, ASCE,* Vol. 106, No. TE1. Also - "ILLI-PAVE PC Version - User's Manual", 1992. Submitted to Transportation Research Board, as a Deliverable for Project NCHRP 1-26.

Research and Development of the Asphalt Institute's Thickness Design Manual (MS-1).1982. Research Report 82-2, 9th Edition, The Asphalt Institute.

Rumelhart D. E., Hinton, G. E., & Williams, R. J., 1986. Learning Representations by Back-Propagating Errors, *Nature*, Vol. 323, pp. 533-536.

Rumelhart, D. E., Hinton, G. E., & McClelland, J. L., 1986. A General Framework for Parallel Distributed Processing, Parallel Distributed Processing: Explorations in the Microstructure of Cognition, Vol. I: Foundations, MIT Press, Cambridge, MA, pp. 45-76.

Southgate, H.F. & Deen, R.C., 1987. Pavement Design Based on Work. Report UKTRP-87-29, Kentucky Transportation Research Program, College of Engineering, University of Kentucky, Lexington.

Specht, D. F., 1991. A General Regression Neural Network, *IEEE Transactions on Neural Networks*, Vol. 2, No. 6, pp. 568-576.

Tabatabaie, A.M. & Barenberg, E.J., September 1980. Structural Analysis of Concrete Pavement Systems. *Transportation Engineering Journal, ASCE*, Vol. 106, No. TE5,

Toll, D., October 1996. Artificial Intelligence Applications in Geotechnical Engineering. Electronic Journal of Geotechnical Engineering, (http://geotech.civen.okstate.edu/ejge/), Vol. 1.

Tutumluer, E. 1995. Predicting Behavior of Flexible Pavements with Granular Bases. Ph.D. Dissertation for the School of Civil and Environmental Engineering, Georgia Institute of Technology, Atlanta GA, USA.

Tutumluer, E. & Meier, R. W., 1996. An Attempt at Resilient Modulus Modeling Using Artificial Neural Networks, *Transportation Research Record 1540*, TRB, National Research Council, Washington, D. C., pp. 1-6.

Tutumluer, E. & Seyhan, U., January 1998. Neural Network Modeling of Anisotropic Aggregate Behavior From Repeated Load Triaxial Tests, Preprint No. 980504, 77th Annual Meeting of the Transportation Research Board, Washington, D.C.

Van Cauwelaert, F. J., Delaunois, F., & Beaudoint, L., ,1986. Computer Programs for the Determination of Stresses and Displacements in Four-Layered Systems, WES Research Contract DAJA45-86-M-0483, U. S. Army Engineer Waterways Experiment Station, Vicksburg, MS.

Wang, K. C. P., 1995. Feasibility of Applying Embedded Neural Net Chip to Improve Pavement Surface Image Processing, *Journal of Computing in Civil Engineering, ASCE*, Vol. 1, , pp. 589-595.

Wasserman, P. D., 1993. Advanced Methods in Neural Computing, Van Nostrand Reinhold, New York,.

Werbos, P. , 1974. Beyond Regression: New Tools for Prediction and Analysis in the Behavioral Sciences, Ph.D. Dissertation, Harvard University.

Williams, T. P. & Gucunski, N. January 1995. Neural Networks for Backcalculation of Moduli from SASW Test, *Journal of Computing in Civil Engineering, ASCE*, Vol. 9, No. 1,pp. 1-8.

Zhu, J-H & Zaman, M. M., January 1997. Neural Network Modeling for a Cohesionless Soil, Preprint No. 970661, 76th Meeting of the Transportation Research Board, Washington, D. C.

Artificial Intelligence and Mathematical Methods in Pavement and Geomechanical Systems, Attoh-Okine (ed.)
© *1998 Balkema, Rotterdam, ISBN 90 5809 028 0*

Evaluation of artificial neural networks for project selection

Gerardo W. Flintsch
The Charles E. Via, Jr. Department of Civil and Environmental Engineering, Virginia Polytechnic Institute and State University, Blacksburg, Va., USA

John P. Zaniewski
Civil and Environmental Engineering Department, West Virginia University, Morgantown, W.Va., USA

Alejandra Medina
Center for Transportation Research, Virginia Polytechnic Institute and State University, Blacksburg, Va., USA

ABSTRACT: This paper presents an application of artificial neural networks to assist decision-makers with project selection-level pavement management decisions. A commercial artificial neural network simulator was trained with historical pavement preservation programming data. The neural network was developed using a designed experiment, and the best trained network was implemented in a computer program that screens the pavement management database and recommends candidate sections for the five-year pavement preservation program. The main conclusion is that artificial neural networks can be useful to assist decision-makers with the selection and recommendation of candidate sections if a reliable source of historical data is available. The designed experiment technique was very useful in determining the most appropriate artificial neural network architecture.

1 INTRODUCTION

The U.S. is served by one of the best transportation infrastructures in the world. The roadway infrastructure alone includes nearly 6.3 million kilometers of street, roads, and highways, and more than 570,000 bridges (BTS, 1997). It is a primary responsibility of federal, state, and local highway agencies to maintain this infrastructure at appropriate levels of service and safety with a minimum cost to the taxpayers. With a deteriorated infrastructure, increasing traffic demands, and shrinking resources, highway agency decision-makers are urged more than ever to optimize the limited resources available for maintaining their assets. Pavement management systems (PMS) play a significant role in the intelligent renewal of the pavement component of the transportation infrastructure by providing a systematic process to handle, analyze, and summarize information for use in selecting and implementing cost-effective pavement construction, rehabilitation, and maintenance programs.

1.1. *Pavement Management Levels*

Haas *et al.* (1994) identify three levels of pavement management, as illustrated in Figure 1. The *project level* focuses on specific technical concerns regarding individual sections. Therefore, it requires the most detailed information and more complex prediction models. A typical output would be a set of design strategies that minimize total life-cycle cost. The other two levels concentrate on administrative decisions that affect programs and policies for the entire road networks. The *project selection level* involves the identification of what projects should be carried out each year of the programming period. The pavement performance models used are less complex and require less detailed data. The *program level* involves general budget allocation decisions for an entire network, requiring even more general information. The models for program level pavement management should be designed to optimize the use of funds allocated for rehabilitation and maintenance (Haas *et al.*, 1994). These three levels of pavement

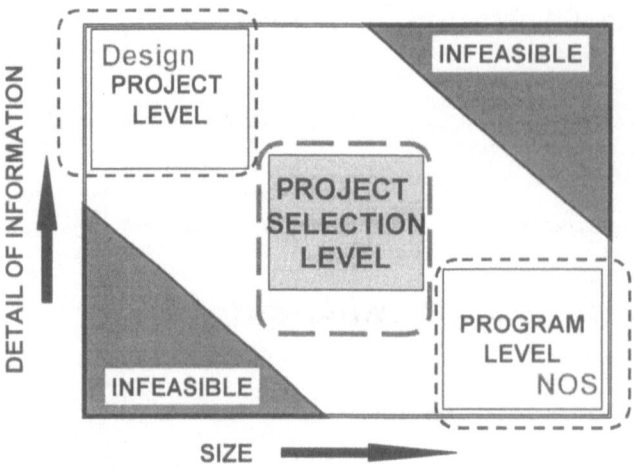

Figure 1. Information Detail and Complexity of Models for a Three-level PMS (after Haas *et al.* 1994)

management can be found in a number of agencies, as is the case in the Arizona Department of Transportation (ADOT).

1.2. *ADOT Pavement Management System*

The Arizona Department of Transportation has been one of the leaders in the implementation of pavement management systems (Kulkarni *et al.*, 1980). The foundation of ADOT's pavement management system is a detailed database that includes data for roughness, cracking, rutting, flushing, patching, skid resistance, structural number (SN), maintenance cost, and traffic volumes and loads for every milepost in the state roadway network. This information is stored by year as shown in Figure 2; a new column for each pavement condition indicator is added each year. A separate database include information about projects, and contains treatments performed and dates at which each treatment was applied.

The program level management decisions are supported by a network optimization system (NOS) that determines the optimum budget needs and allocation for ADOT's pavement network using stochastic analysis and linear programming. The NOS recommends the miles of pavement, in each road functional category, traffic level, and design region, that should receive each rehabilitation action, and the budget required for these actions (Wang *et al.*, 1994). It represents a very effective planning tool for generating annual budget needs.

On the other extreme of Figure 1, the pavement design section handles the project level decisions. After the sections are selected, the most appropriate treatment for each section is designed using life cycle cost analysis.

The pavement management engineer does the project selection. He/she screens the database, selects candidate sections, discusses possible projects with the maintenance districts, and prepares a pavement preservation program. This paper discusses an attempt to use artificial neural networks to assist with these decisions. The artificial neural network was used to "learn" the knowledge applied in the past. The trained network is used to screen the pavement sections in the database and recommend candidate sections for the pavement preservation program. This effort is part of a larger scale research project to develop and implement an automatic project recommendation extension of ADOT's pavement management system (Flintsch and Zaniewski, 1997).

1.3 Artificial Neural Networks and Infrastructure Management

Artificial neural networks are models structured upon the organization of a human brain that can

14

FIELD NAME	A	RTNO	D	MP	L	PW	LS	RS	DIST	AREA	REG	ADT	ADL	GF	->
	I	8	E	1	2	24	5	10	Y	7	0.4	11915	1247	4.3	->
	I	8	E	2	2	24	4	10	Y	7	.0.4	11915	1247	4.3	->
DATA RECORDS	.	.	.	.	.	.	.	.	.	.	. .	.	.	.	->
	.	.	.	.	.	.	.	.	.	.	.	.	.	.	
	U	191Y	N	89	2	24	5	5	S	4	1.9	167	16	1.0	->
	U	191Y	N	90	2	24	5	5	S	4	1.9	167	16	1.0	->

C79	...	C94	R72	...	R94	M72	...	M94	RUT86	...	RUT94	P79	...	P94	F79	...	F94	->
00	...	05		...	49		...	62	0.12	...	0.20	00	...	00	5.0	...	4.0	->
00	...	15		...	57		...	45	0.29	...	0.30	00	...	00	4.5	...	4.0	->
.	.	.	.	.	.	.	.	.	.	.	.	.	.	.	.	.	.	->
.	.	.	.	.	.	.	.	.	.	.	.	.	.	.	.	.	.	
05	...	12		...	112		...			...	0.20	00	...	00	5.0	...	4.0	->
03	...	00		...	123		...			...	0.15	00	...	00	5.0	...	4.0	->

MC_79	...	MC_94	SN	PROJECT	YR	L1	T1	L2	T2	...	L6	T6	T	RATE	FY
4	...	101	3.00	IIG 8- 1- 62	77	SM	12.0	AC	4.5	...			F	16.7	
1412	...	3202	3.00	IIG 8- 1- 62	77	SM	12.0	AC	4.5	...			F	30.3	
.	.	.	.	.	.	.	.	.	.	.	.	.	.	.	
.	.	.	.	.	.	.	.	.	.	.	.	.	.	.	
0	...	0	2.96	S207- - 12	61	SM	18.0	AB	3.0	...			F	28.8	
0	...	0	2.96	S207- - 12	61	SM	18.0	AB	3.0	...			F	77.0	

Figure 2. Data Format of the Pavement Management Database for ADOT (after Wang *et al.*, 1993).

Note: ->: Continue.
Field Name Designations:
   A: Functional Classification, I-Interstate, S = State Highway, U = US Highway;
   RTNO, D, MP, L: Route Number, Direction, Milepost and Number of Lanes, respectively;
   PW, LS, RS: Pavement Width, Left Shoulder Width, and Right Shoulder Width, respectively;
   DIST, AREA, REG: District, Engineering Area, and Regional Factor, respectively;
   ADT, ADL, GF: Average Daily Traffic, Average Daily Load (ESAL's), and Growth Factor, respectively;
   $C_i, R_i, M_i, RUT_i, P_i, F_i, M\_C_i$: Cracking, Roughness, Mu Meter Number, Rutting, Patching, Flushing and Maintenance Cost for year i, respectively;
   SN, Project, Yr: AASHTO Structural Number, Project Number, and Year of Last Project Applied;
   L_i, T_i, Rate, Fy: Type and Thickness of Layer i of Last Project, Priority Number and Fiscal Year.

learn if provided with a range of examples, can deduct their own rules for solving problems, and can produce valid answers from noisy data (Taylor, 1995).

Their architecture, as shown in Figure 3, is characterized by a large number of simple neuron-like processing units interconnected by a large number of connections. These artificial neurons maintain only one piece of information or level of activation, and are capable of simple computations: they receive input from their neighbors, modify their state of activation, a_i, compute an output, o_i, and send that output to their neighbors. The pattern of connectivity among the processing units and the strength of the connections encode the knowledge of a network.

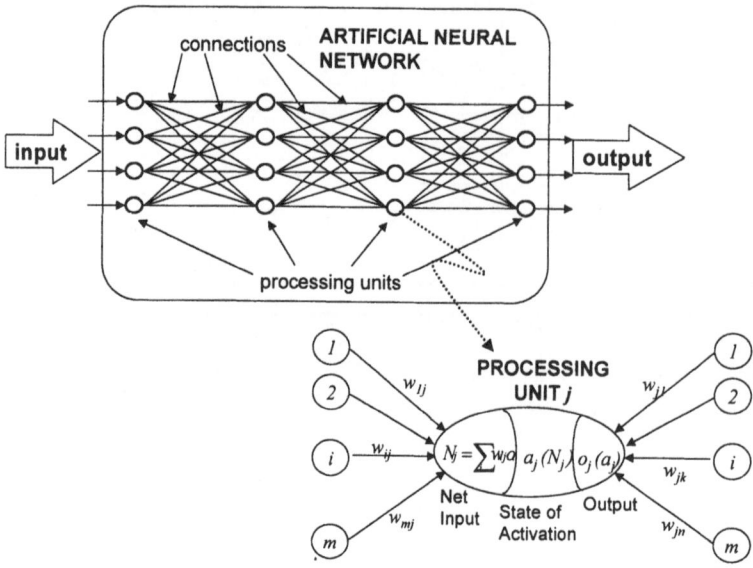

Figure 3. Sample Neural Network (after Garrett *et al.*, 1990).

The main characteristic of artificial neural networks is that they are capable of self-organizing and learning. They are not programmed in the classical sense, but rather they are "trained." A series of examples of the concept to be captured is presented to the system and the network internally organizes itself to be able to reconstruct these examples. The learning can occur in a supervised or unsupervised fashion. For supervised learning, the output expected from the network is included in what the network is to learn. For unsupervised learning, the network is not told what it is supposed to learn from the input presented, and must discover by itself regularities and similarities among the input patterns.

Artificial neural networks are particularly appropriate to solve associative-type problems in which the rules of the system under study that link the input with the output are not clearly defined, but there are a large number of examples over a wide range of input variables. In these cases, an appropriately trained network will perform the task to a suitable level of efficiency. Neural networks have been used for asset condition assessment (Pant *et al.*, 1993, and Van der Gryp *et al.*, 1998), performance prediction (Attoh-Okine, 1995, and La Torre *et al.*, 1998), project selection (Hajek and Hurdal, 1993), prioritization (Fwa and Chan, 1993), and resource optimization (Razaqpur *et al.*, 1996). All of these examples used feedforward neural networks trained using supervised training before implementation. Hybrid systems, which combine neural networks with other artificial intelligence techniques, have been used for automatic analysis of digital pavement images (Ritchie *et al.*, 1991), interpretation of results obtained form nondestructive techniques for condition assessment (Martinelli *et al.*, 1995), and project selection (Taha and Hanna, 1995).

2   ARTIFICIAL NEURAL NETWORKS FOR PROJECT SELECTION

Artificial neural networks were identified as an appealing technique for screening and recommending roadway sections for the pavement preservation program because the process is relatively unstructured, the degree of detail of available information varies from section to section, and the decisions are based on uncertain and sometimes incomplete data.

A commercial back-propagation neural network simulator, BrainMaker (CSS, 1993), was trained with examples of sections programmed and not programmed from past pavement preservation programs. The most appropriate network architecture and training parameters were then defined using a designed experiment. The best-trained artificial neural network was implemented in a computer program that screens and recommends candidate sections when supplied with the current condition of the pavement sections in the network. These candidate roadway sections are further analyzed to prepare the list of sections recommended for the pavement preservation program. In addition to the sections recommended by the artificial neural network, all sections with any of the performance parameters exceeding an acceptable threshold value, or with very low remaining service life, are included in the list of candidate sections.

The design of the artificial neural network included basically two steps: data preparation and selection of input and output parameters, and selection of the most appropriate network architecture and training parameters.

2.1   Data Preparation and Parameter Selection

The Pavement Preservation Programs from 1990 to 1996 were used to prepare the examples for training and testing the neural network. Although older data were available, it was felt that the selection criteria and/or the roadway conditions could have changed significantly.

The average pavement condition, traffic, and maintenance costs were compiled for each section. If only a portion of the mileposts in the section had data, the average of the available points was assigned to the whole section. For four-lane divided highways, the averages were computed for each direction, and the worst of both directions was considered for each parameter. The following information was compiled for each section:

- Classification (interstate or non-interstate)
- Geographic Region
- Structural Number
- Cracking (% of the pavement surface; at the programming year and 1, 2, and 3 years before)
- Roughness (inches/mile; at the programming year and 1, 2, and 3 years before)
- Skid Resistance (Mu-meter Number)
- Rutting (inches)
- Patching (% of Pav. Surface)
- Flushing
- Average Daily Traffic (ADT; vehicles per day)
- Maintenance Cost (dollars)
- Rate, a prioritization index that is computed as a linear combination of cracking, roughness, rutting, and maintenance cost

Since the artificial neural network had to "learn" to differentiate sections that required preservation from those that did not, an equal number of sections non-programmed was selected for each programming year. An output variable *prog* was added to the database to indicate whether the section was programmed. The variable was assigned a value of 1 if it the section was programmed and 0 if it was not.

A total of 418 examples were prepared. Seventy-five percent of these examples were used for training, and 25% were reserved for testing. The training and testing examples were separated randomly.

2.2   Network Architecture and Training Parameters

A designed experiment was used to develop the network and select the most appropriate training parameters. The objectives of the experiment were to identify the neural network design factors that significantly affect the network performance, and the levels at which these factors should be used.

After reviewing the literature available (Lawrence, 1994; CSS, 1993, Maren *et al.*, 1990), four factors were selected for the experiment:

- learning constant or learning rate,
- training tolerance,

□ stratification of the input data, and
□ number of neurons in the hidden layer.

The learning rate is the factor used to scale the corrections done to the weights while the network is learning. Since the learning rate affects the convergence of the network, it could affect the network performance. The training tolerance is the maximum deviation from the actual output accepted to consider a network output correct. The number of hidden neurons, or neurons in the hidden layer (only one hidden layer was used), normally affect the performance of a neural network. The possibility of stratifying or dividing the input fields into ranges was studied because most neural networks respond better to ranges in input or output than to precise numeric values (CSS, 1993). Two separate sets of training and testing examples were prepared. In the first set, the input variables were the numeric values of each field as they were computed (e.g., 5 % cracking). In the second set, the input was coded or stratified by classifying each input item into a maximum of five ranges or categories (low, medium or average, high, very high, and extremely high).

Lawrence (1994) provides several rules of thumb that can be used for the artificial neural network design. According to these suggestions, some two-factor interaction could be expected to be significant. Consequently, a resolution IV, one-quarter fractional factorial experiment (2^{4-1}) was designed using the alias structure suggested by Montgomery (1991). This experiment design permits evaluating 4 factors in 8 runs, with no two-factor interaction aliased with the main effects. Table 1 summarizes the setting selected for the experimental runs.

Each experimental run consisted of a separate training section that started with a complete randomized set of connections between neurons. All training or learning sections were limited to a maximum of 10,000 epochs, and were set to automatically save every 200 training epochs, and to test every 5 epochs. An epoch is completed when all training examples have been presented to the network that has made the corresponding adjustments. BrainMaker's default activation function was used.

Since there is no unique way to measure the performance of an artificial neural network, two sets of networks were analyzed, and several performance statistics were used as response variables in each of them. Table 2 shows the statistics for the network that learned the most training examples in each experimental run. It shows the number of epochs required to reach the indicated level of training, the percentage of examples learned, the percentage of right answers with a tolerance of 0.5, the mean square error of the prediction for all training examples, and the percentage of correct answers, with a tolerance of 0.5, when using the test examples.

The percentage of examples correctly identified with a tolerance of 0.5 was included in the analysis because it could happen that the runs with the higher tolerance have more examples learned, just because of the looser tolerance. The 0.5 value was selected because it is the middle value of the range of the predicted variable (from 0 to 1). Using this tolerance, those sections with a *prog* value larger than 0.5 should be programmed, and those with a *prog* value lower than 0.5 should not be included in the program.

Table 3 shows the statistics for the best testing network in each experimental run. It shows the number of epochs at which the best testing network was obtained, the percentage of testing

Table 1. Selected Levels for the Fractional Experiment Designed.

Training Session	Learning Rate	Training Tolerance	Stratified Input	Hidden Neurons
1	0.2	0.1	Yes	12
2	1	0.4	Yes	12
3	0.2	0.4	No	12
4	1	0.1	No	12
5	0.2	0.4	Yes	48
6	1	0.1	Yes	48
7	0.2	0.1	No	48
8	1	0.4	No	48

Table 2. Best Training Network in Each Experimental Run.

Experimental Run	At Epoch Number	Examples Learned (%)	% Right Train Set	MSE Train Set	% Right Test Set
1	7,400	63%	93%	0.205	62%
2	7,200	88%	89%	0.298	63%
3	6,800	68%	71%	0.367	56%
4	6,400	59%	60%	0.355	58%
5	6,600	99%	99%	0.080	63%
6	3,000	99%	99%	0.075	62%
7	7,400	61%	82%	0.178	62%
8	9,600	82%	82%	0.319	61%

Table 3. Best Testing Network in Each Experimental Run.

Experimental Run	At Epoch Number	% Right Test Set	MSE Test Set
1	7730	66%	0.555
2	6380	68%	0.498
3	715	58%	0.476
4	7980	62%	0.589
5	565	65%	0.534
6	1240	66%	0.542
7	740	68%	0.505
8	5365	66%	0.505

examples correctly identified, and the mean square error of the predictions for all testing examples. It must be noted that the networks that correctly classified the most examples of the training set do not agree with those that have the highest number of right predictions using the test set. The most appropriate network architecture would be the one that maximizes the number of right answers of both the training and testing sets.

The statistical significance of the effects of the various factors considered on the performance measurements was analyzed using Analysis of Variance. The analysis showed that the stratification of the data and the number of hidden neurons have a significant effect on the number of examples learned and the MSE for the training set. Learning rate and training tolerance have no significant effect on training. At the levels used in the experiment, none of the factors have a significant impact on the testing performance. A network with stratified input and a high number of hidden neurons would learn more examples with a lower MSE. The analysis results also suggested that, although training tolerance and learning rate were not statistically significant, setting these parameters at their lower levels would probably have more positive than negative effects.

The final setting for the significant parameters was determined by trying networks with variable numbers of hidden neurons and stratification schemes with different numbers of levels and breakdown values. Examples from two more pavement preservation programs, 1997 and 1998, were added at this time. Figure 4 shows a plot of the number of testing examples correctly identified as a function of the number of hidden neurons. A network with 27 hidden neurons had the best performance. The performance degradation with the higher number of hidden neurons could be due to the fact that the network starts to memorize the examples, losing some of its ability to generalize.

The network with the best testing performance was selected for implementation. This network learned 67% of the training facts with a training tolerance of 0.1, correctly classified

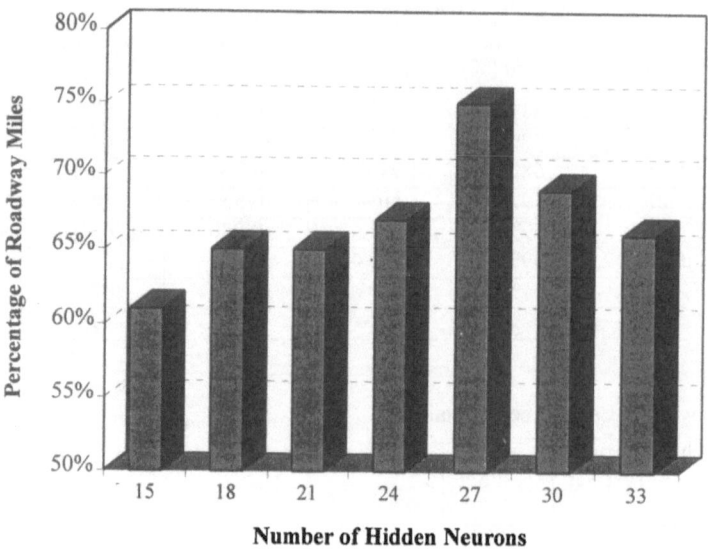

Figure 4. Number of Testing Examples Correctly Identified.

99% of the training examples with a testing tolerance of 0.5, and predicted a correct output for 75% of the testing examples with a testing tolerance of 0.5. This performance was considered acceptable; especially if we consider that not all the reasons for selecting projects for the pavement preservation program are directly inherited from the information in the PMS database. Engineers doing project selection always consider other factors, which are not available in the database, such as a combination of sections due to their geographical proximity, budgetary constraints, distribution of funds among districts, etc. Thus, a "perfect" automatic project selection system is probably impossible.

3   IMPLEMENTATION

To facilitate the day-to-day use of the system, the simulation of the artificial neural network was implemented through a computer program. Figure 5 shows the scheme for the implemented artificial network simulator. For each pavement section, the input processor computes the input values for each field and passes them to the artificial neural network simulator. The simulator uses the neural network architecture defined and computes a *prog* number for each section in the network. An output processor translates this *prog* value to a recommendation. The simulation is repeated for each uniform pavement section in the database to prepare the recommended list of sections for the preservation program. This program is part of a software package that further analyzes these sections.

Using the 1994 data, the program recommended 671 homogenous sections for further evaluation. The average section length was approximately 6.5 kilometers (4 miles). The number of kilometers and sections recommended was higher than the number of kilometers and sections programmed each year. The candidate sections in this preliminary list are then prioritized and a smaller subset of projects is selected for further consideration.

Based on the experience collected from this experiment and taking into consideration the dynamic nature of the pavement management process, the authors believe that a neural network that could learn as it is used will perform significantly better. The dynamic learning capability of the system will be achieved through a hybrid system that combines neural networks with fuzzy logic, genetic algorithms, machine learning, and expert systems. Artificial neural networks will be the backbone of the hybrid system because of their ability to learn from

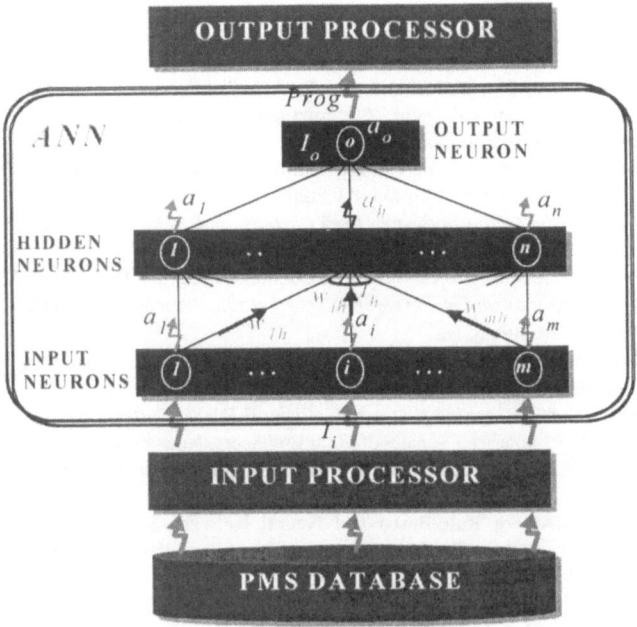

Figure 5. Scheme of the Artificial Neural Network Developed

examples. They will also bring to the table excellent pattern recognition capabilities. Expert systems will contribute their capacity to efficiently handle rules and explanation capabilities. Fuzzy logic concepts will be used for interpretation of the information input and output from the system. Machine learning and genetic algorithms will help with the interpretation of the feedback information and adjustment of the system with time.

4   CONCLUSIONS AND RECOMMENDATIONS

The main conclusion of the research effort is that artificial neural networks can be useful for project selection-level pavement management. This technology can assist decision-makers with the selection and recommendation of candidate sections if a reliable source of historical data is available.

The designed experiment technique was very useful in determining the most appropriate artificial neural network architecture. Two design factors, number of hidden neurons and stratification of the input, were found to have a statistically significant impact on the network performance.

The artificial neural network was able to correctly classify most of the training examples, but only 75% of the testing examples. This performance was considered acceptable considering that not all the reasons for selecting the projects for the pavement preservation program are directly inherited from the information in the PMS database.

Based on the experience collected from this experiment and taking into consideration the dynamic nature of the pavement management process, a neural network that could learn as it is used will probably perform significantly better.

ACKNOWLEDGMENTS

The authors are grateful to James Delton, ADOT Pavement Management Engineer, and Larry Scofield, Manager of the Arizona Transportation Research Center, for their advice and support throughout the research.

REFERENCES

Attoh-Okine, N.O., 1995, Analysis of Learning Rate and Momentum Term in Backpropagation Neural Network Algorithm Trained to Predict Pavement Performance, *Preprint for the 74th Transportation Research Board Meeting*, Washington DC.

BTS, 1997, Chapter 1. The Transportation System, *National Transportation Statistics 1998*, Bureau of Transportation Statistics, http://www.bts.gov/btsprod/nts/chp1v.html.

CSS, 1993, *BrainMaker, Neural Network Simulation Software, User's Guide and Reference Manual*, California Scientific Software, Nevada City, CA.

Flintsch, G.W. and Zaniewski, J.P., 1997, Expert Project Recommendation Procedure for ADOT's Pavement Management System, *Transportation Research Record 1592*, Transport Research Board, National Research Council, Washington, DC.

Fwa, T.F. and Chan, W.T., 1993, Priority Rating of Highway Needs by Neural Networks, *Journal of Transportation Engineering*, American Society of Civil Engineers, New York, NY.

Garrett, J.H., Ghaboussi, J., and Wu, X., 1992, Chapter 6 - Neural Networks, *Expert Systems for Civil Engineers: Knowledge Representation*, American Society of Civil Engineers, New York, NY.

Haas R.C., Hudson, W. R., and Zaniewski, J.P., 1994, *Modern Pavement Management*, Kreiger, Melborne, FL.

Hajek, J.J. and B. Hurdal, 1993, Comparison of Rule-Based and Neural Network Solutions for a Structured Selection Problem, *Transportation Research Record 1399*, Transportation Research Board, National Research Council, Washington DC.

Kulkarni, R., K. Golabi, F. Finn and E. Alviti, 1980, *Development of a Network Optimization System*, Arizona Department of Transportation, Phoenix, AZ.

La Torre, F., Domenichini, L., and Darter, M.I., 1998, Roughness Prediction Based on the Artificial Neural Network Approach, *Proceedings of the Fourth International Conference on Managing Pavements*, Volume 2, May 1998, Durban, South Africa.

Lawrence, J., 1994, *Introduction to Neural Networks, Design, Theory, and Applications*, California Scientific Software, Nevada City, CA.

Maren, A., C. Harston, and R. Pap, 1990, *Handbook of Neural Computing Applications*, Academic Press, Inc., San Diego California.

Martinelli, D., Shoukry, S.N., and Varadarajan, S.T., 1995, Hybrid Artificial Intelligence Approach to Continuous Bridge Monitoring, *Transportation Research Record 1497*, Transportation Research Board, National Research Council, Washington, DC.

Montgomery, D.C., 1991, *Design and Analysis of Experiments*, John Wiley & Sons.

Pant, D.P., X. Zhou, R.S. Arudi, A. Bodocsi and A.E. Aktan, 1993, Neural-Network-Based Procedure for Condition Assessment of Utility Cuts in Flexible Pavements, *Transportation Research Record 1399*, Transportation Research Board, National Research Council, Washington DC.

Razaqpur, A.G., and El Halim, A.O., and Mohamed, H.A., 1996, Bridge Management by Dynamic Programming and Neural Networks, *Canadian Journal of Civil Engineering*, Volume 5, No. 23, October 1996, Canada.

Ritchie, S.G., Kaseko, M., and Bavarian, B., 1991, Development of an Intelligent System for Automated Pavement Evaluation, *Transportation Research Record 1215*, Transportation Research Board, National Research Council, Washington, DC.

Taha, M.A., and Hanna, A.S., 1995, Evolutionary Neural Network Model for the Selection of Pavement Maintenance Strategy, *Transportation Research Record 1497*, Transportation Research Board, National Research Council, Washington, DC.

Taylor, J.C., 1995, Chapter 1: The Promise of Neural Networks, *Neural Networks*, Alfred Waller Limited, Henley on Thames, United Kingdom.

Van der Gryp, A., Bredenhann, S.J., Henderson, M.G., and Rohde, G.T., 1998, Determining the Visual Condition Index of Flexible Pavements using Artificial Neural Networks, *Proceedings of the Fourth International Conference on Managing Pavements*, Volume 1, Durban, South Africa.

Wang, K.C.P., Zaniewski, J.P., Way, G., and Delton, J., 1993, *Pavement Network Optimization and Implementation*, Arizona Department of Transportation, Phoenix, AZ.

Wang K.C.P., Zaniewski, J.P., and Delton, J., 1994, Analysis of Arizona Department of Transportation's New Pavement Network Optimization System, *Transport Research Record 1455*, Transport Research Board, National Research Council, Washington DC.

Artificial Intelligence and Mathematical Methods in Pavement and Geomechanical Systems, Attoh-Okine (ed.)
© 1998 Balkema, Rotterdam, ISBN 90 5809 028 0

Application of neural networks to modeling thick asphalt pavement performance

S.Owusu-Ababio
Department of Civil and Environmental Engineering, University of Wisconsin, Platteville, Wis., USA

ABSTRACT: Knowledge of the timing and cost for highway surface or structural improvements requires a good assessment of the current and future performance of the highway network. Multiple linear regression (MLR) methods have, historically, been used to model the performance of highway pavements. However, it has been argued that MLR is potentially proned to errors in very complex applications such as pavement performance where non-linearities often occur in the data set. This potential error is due to the difficulty involved in defining "exact" relationships through the transformation of variables to make them linear. An artificial neural network-based model is developed and compared with an MLR model in this study to predict the performance of thick asphalt pavements (surface thickness ≥152.4 mm) in Wisconsin. Based on the results of this study, it is concluded that the artificial neural network model is the better predictor of performance; it outperformed the MLR model by as much as 15% in prediction.

1 INTRODUCTION

Highway maintenance and rehabilitation plans determine not only what type of improvements should be made but also when the improvement should be made. Knowledge of when an improvement should be applied requires good assessment of the current and future conditions of the highway network. This assessment, together with a coherent policy and set of maintenance standards, would aid highway engineers in determining and allocating available funds as well as identifying cost effective solutions, anticipating when necessary expenditures will occur, and also be able to present proposed strategies, plans, and budgets objectively and convincingly to elected officials. In addition, road user costs in terms of fuel consumption, vehicle wear and tear, and delays would be minimized as a result of providing good surface quality at the right place and at the right time.

The concept of performance, as a measure of highway pavement deterioration has been widely analyzed and discussed by various researchers (Paterson 1987, Nunez & Shahin 1986, Garcia & Riggins 1984). Predicting pavement performance involves the ability to model the complex relationship between the indicator of performance and the major factors -traffic loads, environmental, design, construction, and maintenance practices-affecting pavement performance as well as the interactions among these major factors. Historically, model forms for performance prediction have used multiple linear regression (MLR). However, it has strongly been argued that, like other classical statistical techniques for prediction, MLR reaches its limitations in applications in which non-linearities occur in the data set (Murray 1993). Transforming variables to make the data linear theoretically can enable linear regression to be as accurate as any statistical model but achieving that goal in problems of any complexity such as in pavement performance requires skill, insight, and persistence. If non-linearities are not found and fixed, linear regression will do nothing to help. Since all the variables must be understood as an interrelated group, the use of linear regression on complex problems tends to be arduous, expensive, and potentially error-prone.

New modeling techniques including artificial neural networks (ANN) have been used successfully in other fields including economics, medicine, and stock market research to analyze problems involving very complex relationships and are found to significantly outperform classical statistical methods. This paper examines the use of both ANN and MLR methods to predict the performance of thick asphalt pavements (thickness ≥152.4 mm (6 in.)) in Wisconsin for the purpose of determining the timing and type of maintenance to be applied. Such information would provide an appropriate basis for conducting a rational life cycle cost analysis associated with thick asphalt pavements (TAPs).

2   THICK ASPHALT PAVEMENT USAGE IN WISCONSIN

The development of any performance model requires a database containing variables which influence the indicator of performance. Despite the crucial role of TAPs in minimizing the damaging effects of heavy traffic loads on pavement structures, its use and performance characteristics, have not been well documented in Wisconsin. Hence, the database used for this study was developed through a survey of the Wisconsin Department of Transportation (WisDOT) district offices and selected city governments. City governments were selected on the basis of information obtained from the Wisconsin Asphalt Pavement Association (WAPA). The survey indicated that only two WisDOT district offices had some experience with the use of TAPs. Four cities including La Crosse, Brookfield, Kenosha, and Waukesha indicated the use of TAPs. The majority of TAPs under city governments, however, were not designed in accordance with any standard pavement design procedures. Pavement thicknesses were determined on the basis of established city government policies on pavements developed through years of experience and engineering judgment. Some cities selected standard thicknesses based on width of roadway cross-section while others used major traffic carrier routes, including downtown streets and industrial/business parks as alternative criteria for selecting pavement thickness.

Owing to the lack of use of standard design procedures in the asphalt thickness determination of city pavements, the acquired data were validated in accordance with WisDOT design procedures. Overall, 32 segments out of 66 non-overlaid segments were found to be carrying significantly different traffic loadings than expected on the existing pavement structures. Nine of the 32 segments were found to be over-designed compared to 23 segments which were under-designed. Hence, only 34 pavement segments were finally considered appropriate for inclusion in this analysis. For the purpose of analysis *n*on-*o*verlaid *t*hick *a*sphalt *p*avements (NOTAP) are defined as TAPs which have not received any form of overlay in their life history.

Performance analysis of thick asphalt pavements using regression

The concept of pavement performance as a measure of highway deterioration has been widely analyzed and discussed by many researchers using various performance indicators. The use of combined indices such as the present serviceability index (PSI), pavement condition index (PCI), pavement quality index (PQI), and pavement distress index (PDI) is popular among pavement engineers. Recent modeling techniques, however, are shifting from the combined index approach to a more versatile approach in which major distress modes are individually modeled to better analyze and explain the relationship between distress and pavement serviceability.

The indicator of pavement performance used in this study was the PDI. It ranges from 0 to 100; 0 being the best and 100 being the worst. A zero rating corresponds to a newly constructed pavement with no surface distress while higher PDI values indicate significant pavement surface distresses. A complete description of the PDI concept is presented by WisDOT (1993). The main factors assumed to affect the performance of NOTAP include: the pavement surface thickness (PST), pavement age (Age), traffic level (ESAL/day), base thickness (BT), and road bed condition measured in terms of the soil support value (SSV). Table 1 shows the statistical characteristics of the thirty-four pavements selected for analysis.

Table 1. Statistical characteristics of thick asphalt pavement variables

Variable Statistic	Variable Name					
	Age (yrs)	ESAL/day	SSV	PST(in)	BT (in)	PDI
Mean	13.2	275	5.0	7	6	50
Std. Dev.	5.9	140	0.8	0.6	2.5	23
Range	2-28	23-578	3.5-5.5	6-9	0-12	7-91

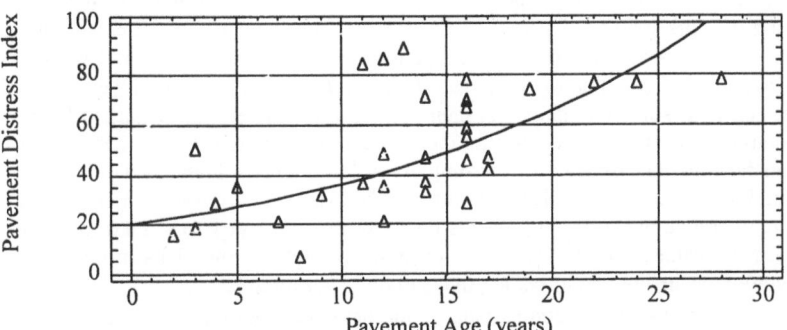

Figure 1: PDI scatter plot and trend for non-overlaid thick asphalt pavements in Wisconsin

Table 2. Performance model for thick asphalt pavements based on regression

Performance Model*	No. of Observations	Model R^2	Model F-value	Model P-value
$PDI(t) = 83.3 + 3.9t + 0.16DESAL - 37.6Log(t \times DESAL)$	34	0.62	15.5	0.0000

*PDI (t) = PDI corresponding to pavement age t
DESAL = Design daily ESALs

The analysis of NOTAP consisted of two phases: a preliminary phase and a model building phase. The former used scatter plots and regression techniques to select key variables for the model building phase. Additional regression analyses were performed in the model-building phase on the variables that were found to influence pavement performance.

Scatter plot and trend line showing the relationship between PDI and pavement age is shown for all NOTAP pavements in Figure 1. Figure 1 shows an increasing trend in PDI with increasing pavement age but in a nonlinear fashion. The regression model and corresponding characteristics pertaining to a confidence level of 95% are summarized in Table 2.

Table 2 indicates that for NOTAPs, only pavement age and traffic loading are significant in explaining the variation in PDI; soil support value is not a crucial variable. This may be due to the fact that, field observations indicated transverse and block cracking as the predominant mode of distress on all of the pavements included in this analysis. Such distresses result from an environmental fatigue process that is determined largely by surface material characteristics and the temperature regime rather than soil conditions. In addition, TAPs, because of their increased structural strength and weight, have the tendency to suppress any roadbed related distress from becoming visible at the surface, thus, having no effect on the PDI (which solely depends on observed surface distress).

3   DESCRIPTION OF NEURAL NETWORK ALGORITHM

Artificial neural networks (ANN) consist of computational models in the form of interconnected nonlinear mesh-like processing elements capable of mimicking neurons, the basic processing elements (PE) of the human brain. The primary concepts and principles concerning the use of ANN has been well documented in the literature (Ian & Nabil 1994, Moselhi *et al.* 1994). A supervised learning backpropagation algorithm is adopted in this study for analysis based on results reported from various fields on the use of backpropagation-oriented neural networks (Attoh-Okine 1994, Prahlad 1994, Zaidi *et al.* 1992). The learning algorithm involves a forward-propagating step followed by a backward-propagating step. Both steps are done for each pattern presentation during network training. During the forward step, the network is presented with an input pattern paired with a target output. The input patterns are presented to the input layer, and continues as activation level calculations propagate forward through the hidden layers. In each successive layer, every PE sums its inputs and then applies a predefined mathematical function to compute its output. The output layer of PEs then produces the output of the network. The mathematical transfer function used by the learning algorithm could be linear or nonlinear; the nonlinear transfer function in this algorithm is the sigmoid (S-shaped curve) which produces values typically in the range of 0 to 1. To use this function, it is necessary to scale or normalize the original data if the values used as inputs are excessively high or low. This minimizes the potential for saturation occurring and impeding the network's ability to learn.

The backward step begins with a comparison of the network's output pattern to the target output by calculating the error or difference. The error values for all hidden PEs and changes for their incoming weights are calculated, beginning with the output layer and moving backward through the hidden layers. In this step the network corrects its weights in such a way as to decrease the observed error. The error value associated with each PE reflects the magnitude of the error associated with that PE. This error is used during the weight-correction procedure, while learning is taking place. A larger value for the error requires a larger correction to be made to the incoming weights, and its sign reflects the direction in which the weights should be changed. Weights associated with each interconnection are adjusted during learning. A set of input/output pattern pairs is used for training, and is presented to the network many times. After training is stopped, the performance of the network is tested with data unknown to the network.

Performance analysis of thick asphalt pavements using artificial neural networks

The network building process involves determining the architecture of the network which include the number of layers and the number of PEs in each layer which would give an optimum performance for the PDI prediction model. The number of PEs of the input and output layers are defined to be two for the input layer (daily ESAL and pavement surface age) and one for the output layer which corresponds to an estimate of the PDI value. To ensure efficient convergence of the network training and the desired performance of the trained network, several parameters must be determined. These parameters include the learning rate, momentum term, the error terminating value and the number of training cycles. The learning rate is a factor that proportions the amount of adjustment applied each time a weight is updated. The magnitude of this parameter determines more or less, the pace of weight adaptation. However, an excessively large coefficient often causes the convergence behavior of the network to oscillate and possibly never converge. A small learning rate usually causes the learning process to progress slowly and has a potential of resulting in local minima conditions. To compensate for the slow weight adaptation when using a small learning coefficient, a momentum term is adopted (Rumelhart *et al.* 1986). This coefficient increases or decreases the weight adjustments by an additional amount based on the direction of the adjustment of the previous iteration. In a study to analyze the effect of

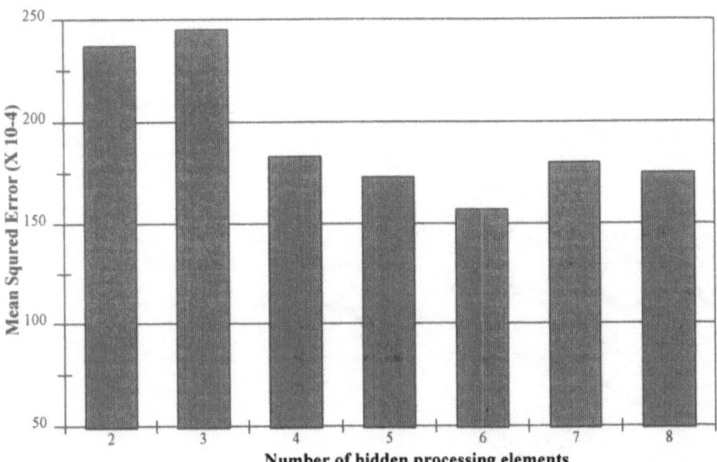

Figure 2: Mean squared error as a function of number of 1-hidden layer processing elements

learning rate and momentum on pavement performance prediction, Attoh-Okine (1995) concluded that a learning rate in the range of 0.2-0.5 and momentum term in the range of 0.4-0.5 will yield stable model performance. Based on this conclusion momentum terms and learning rates in the specified ranges were initially investigated; this resulted in the use of a momentum value of 0.45 and learning rate of 0.5 for this study. The error terminating value and the number of cycles provide means of when to stop training. The error value was set to 0.00001. A preliminary analysis on all networks examined indicated that a maximum of 10,000 cycles was adequate for the training of all feasible networks. The mathematical function used in value transfer to hidden and output layer PEs was the sigmoid transfer function, after the input layer had fanned out the input values to the PEs in the hidden layer. A generalized delta learning rule was employed; in this learning rule weight changes are applied to previous weights after the presentation of each input or pattern.

The objective was to determine the optimum number of hidden PEs for the PDI prediction problem. This required the training and testing of every network topology that would be developed. From the database all the 34 cases or observations (each case consisting of the observed PDI value, pavement surface age in years, and ESAL) were selected for training. Due to the small sample size the network could not be tested with data outside the range of the data set used for training. With one-hidden layer, the objective was to determine by trial and error the number of PEs to yield an optimum and stable model performance. The use of one hidden layer for this analysis is based on conclusions from an earlier study which concluded that a one-hidden layer may be adequate for pavement performance modeling (Owusu-Ababio 1998). A maximum of 8 hidden PEs was investigated. This maximum value is based on a rule of thumb specified by Nelson and Illingworth (1989) which recommends the use of four hidden PEs for every input layer PE , i.e. referring to the number of independent variables, which was two for this analysis. Thus, 7 networks were initially designed and trained until convergence was achieved using hidden PEs in the range of 2 to 8 in increments of 1. Ideally, each network developed would require validation and testing with data not included in the model building process (out-of-sample data). The network that produces the minimum mean squared error on the out-of-sample data would be the preferred network solution. However, in this study, due to the small sample size it was not possible to test the network with an out-of-sample data. Hence, the trained sample data (in-sample data) was used to test the network. The network with the minimum mean squared error (MSE) was selected as the optimum network. The mean squared error is calculated as given in equation 1:

$$MSE = \Sigma(O - t)^2/N \tag{1}$$

where   O = output vector, t = target vector, and N = number of patterns.

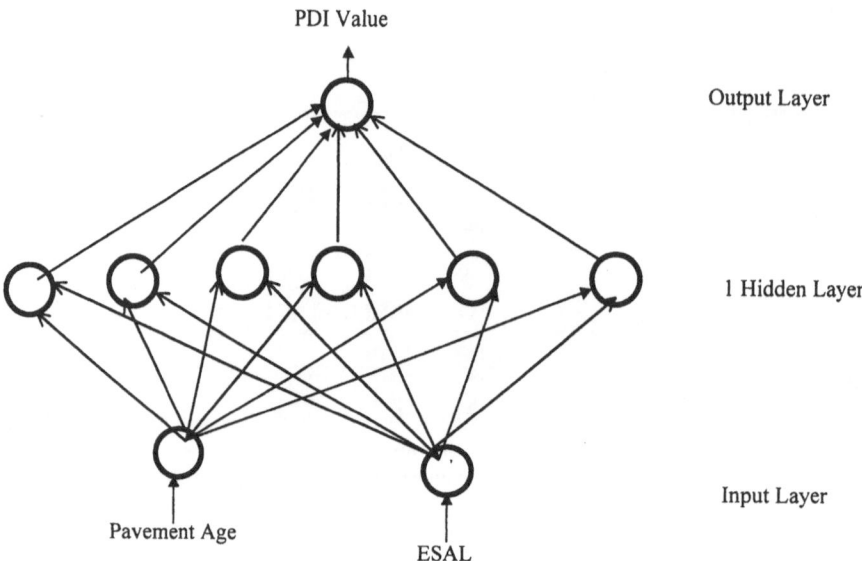

PDI Value

Output Layer

1 Hidden Layer

Input Layer

Pavement Age

ESAL

Figure 3: Backpropagation neural network architecture for the PDI prediction model

The network with the minimum MSE on the in-sample test data was the one with 6 hidden PEs (MSE = 157 x 10^{-4}) as shown in Figure 2. The overall topology for this optimum network is shown in Figure 3. It has a 2 x 6 x 1 (2 by 6 by 1) topology; i.e. 2 input PEs corresponding to the input variables (pavement surface age and ESAL); 6 hidden PEs; and 1 output PE which produces the value of the predicted PDI.

4   COMPARISON OF NEURAL NETWORKS AND REGRESSION MODELS

The fundamental objective of each of the two methods investigated in this study was to fit an accurate model for the universe of the PDI data. The accuracy of fitness typically is measured by parameters such as the mean absolute error (MAE), the standard error of the estimate (SE), and the coefficient of determination (R^2) of predicted versus actual values. The criterion to determine whether the ANN model outperforms the MLR model is, if it converges to smaller SE and has higher R^2 for actual versus predicted data when compared with the MLR model. The SE and R^2 measures are used to compare the two methods based on the in-sample data i.e. data used for developing the models. Similar computations for an out-of-sample data set provides an appropriate basis for judging the ability of each of the two methods to generalize or give a reasonable prediction beyond the range of data set used in their development but in this study it was impossible due to the small sample size. Figure 4 shows scattergrams depicting the amount of predicted versus observed PDI for both ANN and MLR model types. The ideal shape in these scattergrams would be a straight line with a slope of 45 degrees that crosses the origin as shown in Figure 4. The scattergrams, however, show deviations from the ideal for both model types, but it is evident that the ANN model gives the "better" performance for the in-sample test data. These results are reflected in the SE and R^2 values presented in Table 3. Table 3 indicates that the ANN model outperforms the MLR model by 15% based on the R^2 values. Thus 77% of the variation in PDI can be explained by pavement age and traffic loading using ANN as the prediction tool compared to 62% if MLR is used. This is due to the fact that, non-linearities occur in the data set as indicated by the trend line shown in Figure 1. The MLR attempts to transform variables to make the data linear but this transformation process might not have been able to capture the "exact" nonlinear relationship; thus making the MLR approach less accurate compared to the ANN method.

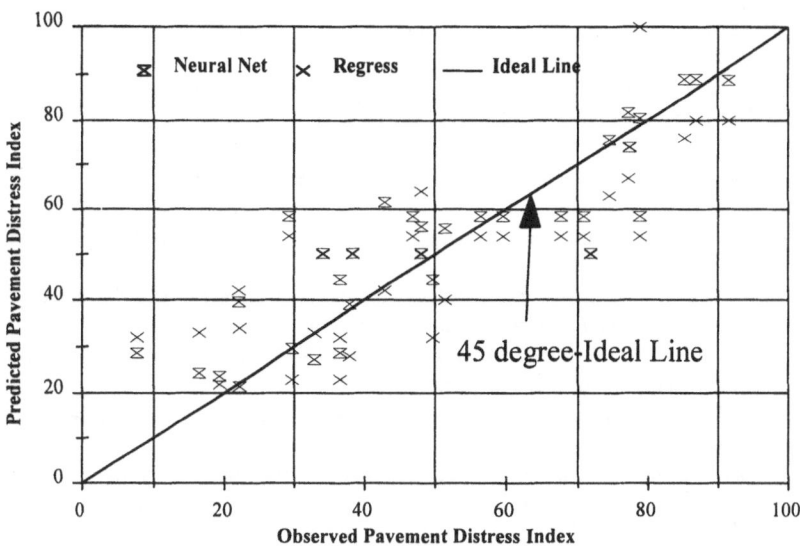

Figure 4: Predicted versus Observed PDI for ANN and MLR Models

Table 3. Comparison of Artificial Neural Networks and Multiple Linear Regression Models

Model Type	Standard error (S.E.)	R^2
ANNs	11.3	0.77
MLR	15.0	0.62

5 SUMMARY AND CONCLUSIONS

This paper examined the development and predictive capabilities of two methods --multiple linear regression and artificial neural networks-- for modeling the performance of thick asphalt pavements (thickness ≥ 152.4 mm (6 in.)) located in Wisconsin. The data for the model building process support the following observations:

  ❑ The artificial neural network model outperformed the multiple linear regression model by 15% in the prediction of pavement distress index (PDI). The results suggest that the potential of artificial neural network as a tool in pavement performance modeling appears excellent and should be explored using more comprehensive databases.

  ❑ Age and traffic loading are the predominant factors influencing the deterioration of non-overlaid thick asphalt pavements in Wisconsin.

REFERENCES

Attoh-Okine, N.O., 1994. Predicting Roughness Progression in Flexible Pavements using Artificial Neural networks. *Proceedings of the Third International Conference on Managing Pavements*, Vol. 1:San Antonio, TX., pp 55-62.

Attoh-Okine, N.O., 1995. Analysis of Learning Rate and Momentum Term in Backpropagation Neural Network Algorithm Trained to Predict Pavement Performance. *Paper presented at the74[th] Annual Transportation Research Board Meeting:* Washington, D.C.

Garcia-Diaz, A., & Riggins, M., 1984. Serviceability and Distress Methodology for Predicting Pavement Performance. *Transportation Research Record* 997: 17-23

Ian, F., & Nabil, K., 1994. Neural Networks in Civil Engineering: I. Principles and Understanding. *Journal of Computing in Civil Engineering*, 8(2): 131-148

Moselhi, O., Hegazy, T. & Fazio, P., 1994. Developing Practical Neural Network Applications using Back-propagation. *Microcomputers in Civil engineering*, 9(2):145-151

Murray, S. 1993. *Neural Networks for Statistical Modeling*. New York: Van Nostrand Reinhold.

Nelson, M.M., & Illingworth, W.T., 1989. A Practical Guide to Neural Nets. Texas Instruments: San Diego, CA.

Nunez, M.M., & Shahin, M.Y., 1986. Pavement Condition Data Analysis and Modeling.*Transportation Research Record* 1070: 125-132

Owusu-Ababio, S. 1998. Effect of Neural Network Topology on Flexible Pavement Cracking Prediction. *Journal of Computer-Aided Civil and Infrastructure Engineering*, Vol. 13: 349- 355

Paterson, W.D.O., 1987. *Road Deterioration and Maintenance Effects: Models for Planning and Management*. Baltimore: John Hopkins University Press.

Prahlad, P., 1994. Neural Network for Gap Acceptance at Stop-Controlled Intersections. *Journal of Transportation Engineering*, 120 (3):432-446

Rumelhart, D.E., Hinton, G.E., & Williams, R.J., 1986. *Learning Internal Representations by Error Propagation- Parallel Distributed Processing: Explorations in the Microtexture of Cognition, Vol. 1*.Cambridge, MA: MIT Press

Wisconsin Department of Transportation, 1993. *Pavement Surface Distress Survey Manual*. Madison, WI: WisDOT-Division of Highways.

Zaidi, A., Refenes, A.N., & Francis, G., 1992. Stock Performance Modeling Using Neural Networks: A Comparative Study with Regression Models. Department of Decision Science, London Business School, Sussex Place, London,

Artificial Intelligence and Mathematical Methods in Pavement and Geomechanical Systems, Attoh-Okine (ed.)
© *1998 Balkema, Rotterdam, ISBN 90 5809 028 0*

Representation of a function with three variables by using neural networks

A. Martin, M. Espina, O. Rodriguez & I. Tansel
Florida International University, Department of Mechanical Engineering, Miami, Fla., USA

ABSTRACT: A back-propagation Neural Network is used to represent a three variable function, in order to investigate the accuracy of this type of System Identification Technique. The selected function generates a three-dimensional surface with several local minimum and maximum, and a well-defined global maximum (pick). After 72 hours, the Neural Network with 20 hidden nodes had the better percentage of total error (0.01149) than the networks with 15 and 25 hidden nodes. Another set of data points was generated for testing purposes and introduced in the back-propagation Neural Network. The output of the system showed values very close to the values of the real function. Approximately a 30% of the output obtained from the testing section had 0% of error and other 30% had a very low percentage of error (0.00833). Just some critical points of the surface received the highest error, but still acceptable (0.025).

1 INTRODUCTION

To emulate human's ability to learn and the human brain's capacity to analyze, reason and, discern represent difficult tasks for computers. The brain, with its millions of neurons is the most perfect and efficient of the computers. Actually, it is a highly complex, nonlinear, and parallel computer. Its capacity to perform certain tasks (e.g. image processing, motor control, pattern recognition, and perception) is several times faster than any available computer.

The sequential, logic-based digital computing method has been successful in many areas. In fact, the computer development has changed the way the humans deal with the future very drastically.

Studies to develop artificial tools, based on mathematical model that simulated the brain performance started about 50 years ago. Currently, as computers are becoming more powerful and versatile, scientists are trying to use the machines more often for tasks that are relatively simple for humans. The Artificial Neural Networks (ANN) is an example of this approach.

2 THEORETICAL ASPECTS

As the name suggests, the Artificial Neural Network is certain information-processing system with performance characteristics similar to biological neural systems. Artificial Neural Networks are based on generalizations of mathematical models from different human behaviors such as human cognition. This is associated with biological structures such as the nervous system.

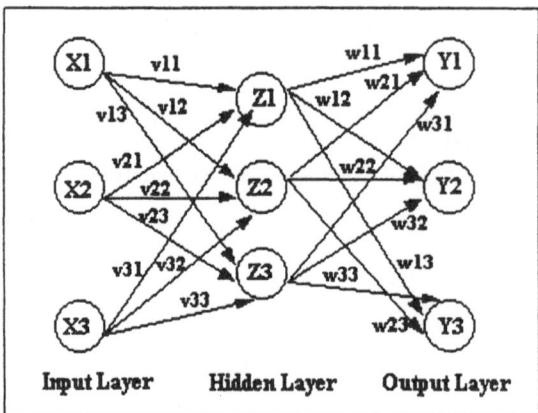

Figure 1. A multi-layer neural network

A neural net consists of a large number of simple processing elements called *neurons or nodes*. The neurons are interconnected to each other by means of *connection links*. These links allow them to exchange information. Each communication link has an associated weight, which represents the information used by ANN to solve the problem. Each neuron has an internal state, called *activation level or function* (often a non-linear function). This is a function of the inputs it has received to determine its output signal.

A Artificial Neural Network is characterized by the pattern of connection between neurons called *architecture*, the method to calculate the weights on the connections called *learning or training algorithm* and the *activation function*.

Neural Networks encompass various techniques for modeling nonlinear processes. They are firmly rooted in statistical methods. They massively use parallel-distributed processors with a natural propensity for storing knowledge based on experience.

The ANN comprises an input layer, a *hidden layer* (frequently more than one) and an output layer, which contains the elements, called *neurons*. Figure1 illustrates a multi-layer Neural Network with one hidden layer.

The limitations of single-layer neural networks are overcome by a new effective general method of training a multi-layer neural network, known as *back-propagation* (of errors) or the generalized delta rule. The back-propagation training method is simply a gradient descent method to minimize the total squared error of the output computed by the net.

The training of a back-propagation net can be divided in three stages:

1. The feed-forward of the input training pattern.
2. The calculation and back-propagation of the associated error.
3. Adjustment of the weights.

During feedforward, each input unit (X0i) receives input and sends the signal to each of the hidden nodes (Zi). The hidden nodes compute their activation and send a signal to each output unit. Then, the output units compute their activation to form the net response for the given pattern.

During training, each output unit compares the computed activation y_k with the target value to determine the error. Based on this error, the factor δ_k is found. This factor is used to distribute the error back to the hidden node units. Similarly, the factor δ_j propagates the error back to the input layer. However, in this case δ_j is used to update the weights between the hidden layer and the input layer. After all of the δ factors have been determined, the weights for all layers are adjusted simultaneously.

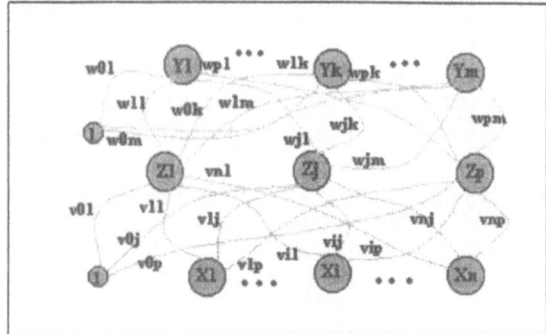

Figure 2. Back-propagation neural network with one hidden layer

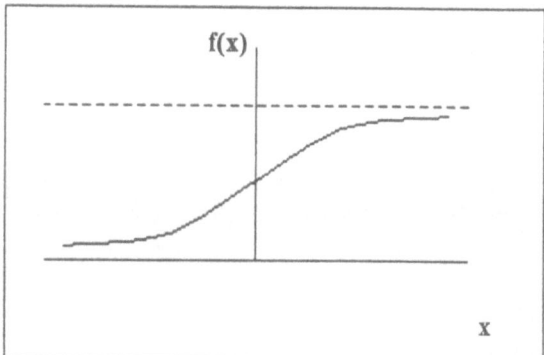

Figure 3. Sigmoid function

A multi-layer back-propagation neural network with one hidden layer is shown in Figure 2.

For back-propagation network, the activation function should be continuos, differentiable, and monotonically non-decreasing. In addition, its derivative should be easy to compute. One of the most typical activation functions is the binary sigmoid function with range between 0 and 1 defined as:

$$f_1(x) = \frac{1}{1+\exp(-x)} \tag{1}$$

Another common activation function is bipolar sigmoid, which has range between −1 and 1, defined as:

$$f_2(x) = \frac{2}{1+\exp(-x)} - 1 \tag{2}$$

Figure 3 illustrates the graphic of the sigmoid function.

Sometimes, convergence is achieved faster if a momentum term is added to the weight update formulas. In order to use momentum, weights from one or more previous training patterns must be saved.

In back-propagation neural networks with momentum, the weight is adjusted in such a way that is a combination of the current gradient and the previous gradient. In doing so, the momentum gives to the net more flexibility to make adequately large weight adjustments. When the net uses momentum, it is not moving in the direction of the gradient, but in the direction of a combination of the current gradient and the previous direction of the weight correction.

Learning rate is a parameter that controls the amount by which weights are changed during training. Changing this parameter during the training process (reducing it as training progresses) can improve the speed for back-propagation, bringing to the system stability.

3 ACCURACY OF BACK-PROPAGATION NEURAL NETWORK

A three-dimensional surface was selected to investigate the accuracy of the back-propagation neural network. The mathematical formula is as follows:

$$Z = \frac{\sin\left(\sqrt{X^2 + Y^2}\right)}{\sqrt{X^2 + Y^2}} \tag{3}$$

The values of X and Y will be the input for the net and the Z values given by the mathematical representation of the function will represent the output. Figure 4 shows a graphical representation of this surface.

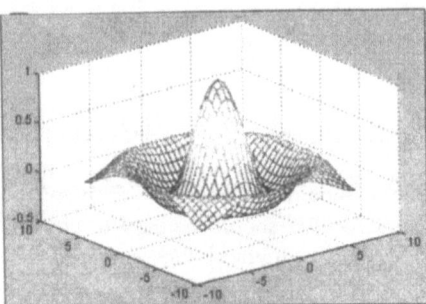

Figure 4. Graphical representation of the 3D function

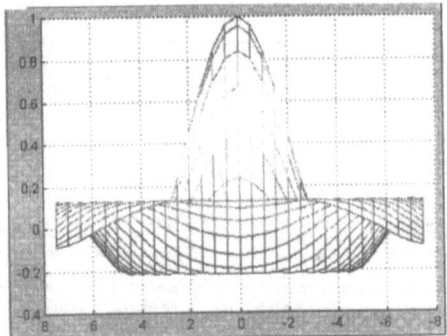

Figure 5. Different viewpoint of the function

This function is undefined for $(X,Y)=(0,0)$, for which the denominator becomes 0. Also the function has a global maximum for values of X and Y near to 0 and several global minimums at Z=-0.2.

Figure 5 illustrates a different view of the surface. With this representation, the boundaries of the function can be established with accuracy.

Form the two-dimensional view below, limits with respect to Z-axis can be obtained using the extreme function values in this direction. The minimum value of −0.2 and the maximum of 1.0 the output range.

The relationship between the input (i.e. X and Y) and the output (i.e. Z) is given by a mathematical equation. The problem arises with the complexity of the surface. Several local minimums and maximums, which are hard to figure out by the neural network are present. Therefore, it is very important how training points are distributed along the surface. Critical points should be included in the training set for better performance.

Points equally distributed along the XY plane covering the entire surface resulted in the best distribution for the training set of points. Figure 6 shows this particular distribution.

The critical points of the data must be given to the neural network to delimit the boundaries of the problem. Table 1 establishes these points for the 3D function.

The data was generated with a simple C program. A double *for* loop guarantees the generation of the array of points. An *if* statement in the inner *for* loop insures avoiding the undefined point.

4 TRAINING PROCESS

During training, the Neural Network a back-propagation neural network called ANALOG was utilized. After the characteristics of the training set have been established, the input file (data for training) is prepared with the set of points. This file has the following format:

total number of entries
input/space/input
output

Table 1. Characteristics of the data.

	Axis	Minimum	Maximum
INPUT	X	-8.00	8.00
	Y	-8.00	8.00
OUTPUT	Z	-0.2	1.00

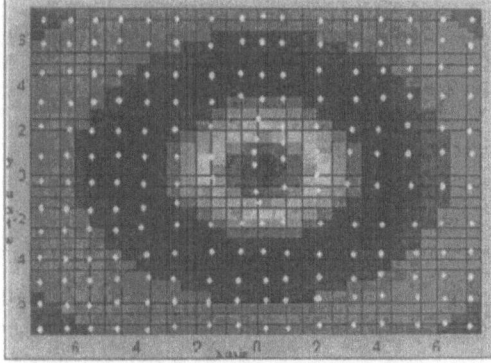

Figure 6. Training points distribution

ANOLOG, by default creates three different types of file (i.e. name.net, name.fig, name.chr) in the training process. After the net has received the data, it performs a duplicate check to eliminate repeated events.

Different neural networks were set up using 15, 20 and 25 hidden nodes. The default values of *learning rate* (0.6) and *momentum* (0.9) were also used. Data access was done randomly by setting *type of presentation* parameter to *random*. This parameter controls the way different events are tested.

After 48 hours, the estimation error had decreased drastically in all neural networks. However, the best statistics were obtained from the neural networks with 20 and 25 *hidden nodes*.

In an attempt to achieve better results, another neural network with 22 *hidden nodes* was developed. Once again, three neural networks were tested but this time the *hidden nodes* were 20, 22 and 25 respectively. After 72 hours of continuous training process, the Learning Error Factors Window in the ANALOG program for the neural network with 22 hidden nodes showed the lowest estimation error. In another 48 hours, the *learning rate* and *momentum* were reduced gradually until it reached the values of 0.2 and 0.6 respectively. The total estimation error was 0.014915. Figure 7 shows the Summary of Learning Error Factors Window for the network with 22 hidden nodes after 5 days of continuous work.

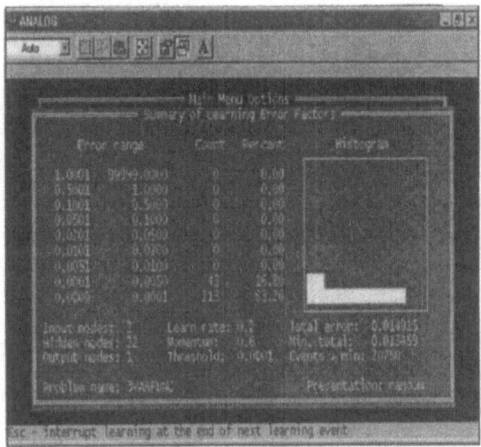

Figure 7. Summary of learning error factors

After these results the training process was interrupted to prepare the testing set.

5 TESTING PROCESS

The next step in the process was to develop a testing data set. Basically, these were intermediate points among those training in the training set. For that purpose, minor changes were made to the C program to adjust the iteration. Another set of data can be obtained by just changing the initialization and condition statements in the double *for* loop. Also, in order to test the reliability of the system the critical point $(X, Y) = (0.01, 0.01)$ with $Z = 0.999 \approx 1.0$ was included.

Next, the test data set file was obtained. Automatically the neural network changes the output and introduces its own results to the system. These results can be compared with the real output of the function to obtain the estimation error.

Table 2. Testing Data Analysis

X	Y	Zreal	Znn	Error
-5.00	-7.00	0.090	0.080	-0.00833
-7.00	-4.00	0.120	0.100	-0.01667
-5.00	-3.00	-0.070	-0.100	-0.02500
-2.00	-1.00	0.350	0.350	0.00000
4.00	0.00	-0.190	-0.150	0.03333
0.00	2.00	0.450	0.470	0.01667
3.00	3.00	-0.210	-0.180	0.02500
-4.00	6.00	0.110	0.120	0.00833
4.00	6.00	0.110	0.110	0.00000
0.01	0.01	0.990	0.940	-0.04167

The formula to calculate the estimation error will be:
Error = (estimated – real) / range
where:

       estimated: neural network output
       real: real output
       range: |min – max|

Table 2 analyzes the estimation error for some testing points in the set. It also shows the maximum and minimum errors obtained.

6 ERROR ANALYSIS

Obviously, the maximum error was obtained for the point nearest to the undefined point $(X,Y)=(0.01,0.01)$ with $Z=0.999 \approx 1.0$ with an error of -0.041667, but it was shorter enough to be considered acceptable. The error was distributed in the following manner:

-    30 % of the data with |0.008333|
-    28 % of the data with |0.016667|
-    23 % of the data with |0.000000|
-    13 % of the data with |0.025000|
-     9 % of the data with |0.033333|

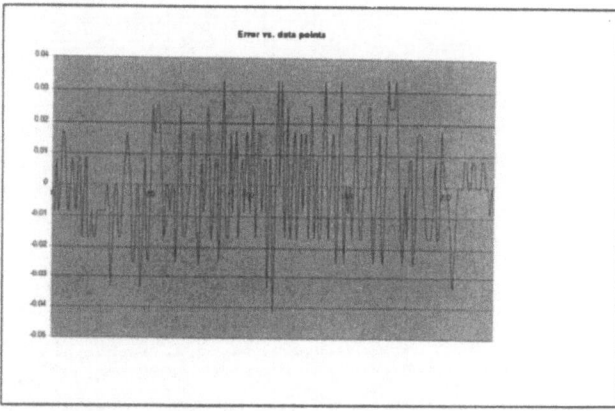

Figure 8. Error distribution for the testing set

Another important issue is that the error was distributed uniformly along the surface. Figure 8 shows a graphic of error vs. testing points in which it can be observed how the error is moving from positive to negative in the same range.

If the percentage of error is analyzed in detail, it is possible to notice that the neural network was very accurate in points distributed along the whole surface. However, the highest errors are found in the peak zone (the middle of the graphic). However, the outer part of the surface has smallest errors.

7 CONCLUSION

This work is another demonstration of how the neural networks can be utilized as a useful tool in system identification. The application of a back-propagation neural network to a complex surface confirmed the capability of this identification technique to "learn" after a short period of training.

The relationship between the data, the boundaries of the training set, the number of hidden layers and the set up of important parameters of the net are the most important issues to take into account when a neural network is going to be used. Parameters such as learning rate, momentum and data presentation are also important when setting up an efficient and accurate neural network.

In this case the neural network was tested with a complex 3D surface given by a mathematical function. Results corroborated that it is possible to apply this technique to this kind of problem with excellent results.

8 REFERENCES

Laurene, F. 1994.*Fundamental of Neural Networks: architectures, algorithms, and applications.*
Obermeier, K.K. & Barron, J.J. 1989. *Time to get fired up*, In depth Neural Networks, Byte.
Richard, P. October 1997. *System developers apply neural networks to fingerprint analysis*, Spotlight on advanced technology, Vision Systems Design.
Simon, H. 1994. *Neural Networks: A comprehensive foundation.*

Artificial Intelligence and Mathematical Methods in Pavement and Geomechanical Systems, Attoh-Okine (ed.)
© *1998 Balkema, Rotterdam, ISBN 90 5809 028 0*

A neural network model for a cohesionless soil

Musharraf Zaman & Jian-Hua Zhu
School of Civil Engineering and Environmental Science, The University of Oklahoma, Norman, Okla., USA

ABSTRACT: A framework of recurrent neural network (RNN) is used to model nonlinear behavior of a cohesionless soil. The RNN model is a dynamic neural system which appears effective in the input-output modeling of complex soil behavior. A dynamic gradient descent learning algorithm is used to train the network, producing substantial improvements between observation and modeling results as compared with a traditional constitutive modeling technique. In developing a traditional constitutive model, assumptions concerning yield/failure state of a soil form a basis of the modeling, and determination of the associated material parameters constitutes a major task of the implementation process. In contrast, the RNN model is based only on the experimental data rather than various assumptions and there is no need to find any parameters that are critical to the traditional constitutive models. Application of the RNN model on a series of triaxial compression shear tests performed on a dune sand illustrates high efficiency of the RNN model. Excellent agreements between the experimental data and modeling results are observed for both stress-strain relationships including unloading and reloading response. The volumetric response in the form of either compression or dilation of the soil during shearing at different stress level are simulated in a quite satisfactory manner. Also, rapid variations in volumetric strains at unloading and reloading stages are well represented by the RNN model.

1 INTRODUCTION

Modeling of nonlinear behavior of cohesionless soils is important in solving many problems in geotechnical engineering. Many research works have focused on this subject during the last several decades (Desai 1990; Duncan & Chang 1970; DiMaggio & Sandler 1971; Drucker *et al*. 1957; Faruque & Zaman, in press; Desai & Watagala 1993; Lade 1977). Most models developed in the past were based on various theories including elasticity, and plasticity (Desai 1990; Duncan & Chang 1970; DiMaggio & Sandler 1971; Faruque & Zaman, in press), which are called here as traditional constitutive models. To develop a traditional constitutive model, one must follow such unified procedures as follows:

(i)   making simplified assumptions that form a basis of the models;
(ii)   proposing criteria for identifying yielding and/or failure state of the soil;
(iii)   formulating specific mathematical expressions consistent with (i) and (i i);
(iv)   finding appropriate material parameters used for back prediction of the soil constitutive behavior (Fukushima, & Tatsuoka 1984; Abduljauward *et al*. 1992; Loung 1980; Azeemuddin 1988).

In general, capabilities of a traditional constitutive model is associated with the complexities of the model (Azeemuddin 1988; Faruque 1987; Zaman *et al*. 1996; Zaman *et al*. 1982). A more complex model is expected to yield more accurate results, but at the same time it requires more parameters to be used in the model (Desai 1990; Duncan & Chang 1970; Azeemuddin, 1988; Zaman, et al. 1996). As it is evident from the literature, the traditional constitutive models have been used for solving many engineering problems and these models have contributed significantly to soil mechanics, foundation engineering and other disciplines. However, the procedures used in implementing some of these constitutive models can be fairly complex. Also, with the development of more complex models, the deficiency in satisfying conservation law of energy and invariance requirements has increased significantly (Ghaboussi, 1992). In addition, evaluation of material parameters in many traditional constitutive models can be an extremely difficult task, requiring specialized tests and proper optimization technique to obtain a unique set of parameters from a given set of experimental data.

Furthermore, many traditional constitutive models generally concentrate on the simulation of stress-strain relationship with less attention given to the volume change behavior of soil, which is an important characteristic of cohesionless soils. The volume response in the form of dilation and compaction of a cohesionless soil is primarily dependent upon the relative density and magnitude of confining stress during the shearing process (Desai & Watagala 1993; Lade 1977; Fukushima, & Tatsuoka 1984; Azeemuddin, 1988; Zaman et al. 1996). The dilative volume change generally occurs in the case of low confining stress for a dense sand, while the contraction is common for loose sand or when confining stress is high for dense sand. These two types of volume response can also be observed in one test conducted at a relatively high range of shear stress. As such, to predict the volume change of the soil accurately one must have a constitutive model that can account for both the compression state and the dilation state of a cohesionless soil and appropriate parameters for these states, thus increasing the total number of parameters required in the model (Azeemuddin 1988; Faruque 1987; Zaman et al. 1996). This, in turn, makes it more difficult to use such models in engineering practice.

An artificial neural network (ANN) offers an alternative approach to model soil behavior. The ANN model is fundamentally different from a conventional constitutive model. One of its distinctive attributes is that it is based on the experimental data rather than mathematical functions and assumptions. This model learns from the experimental data and forms neural connection stimuli from learning process, functioning somewhat like a human brain (Ghaboussi. 1992; Ghaboussi et al. 1991; Elman 1990). When using such modeling technique, a highly complex, but very accurate associations can be directly achieved from the learning process involving experimental data, without developing any mathematical expressions and making any assumptions regarding yield/failure states of the soil (Ghaboussi. 1992). Because of its learning, training and predicting characteristics, the ANN model has a great potential for application in geotechnical engineering where the experimental data available in the literature are rich and there still exists strong needs for an appropriate model, particularly for cohesionless soils (Ghaboussi. 1992b; Ghaboussi et al. 1991).

The available references are quite few with regard to the neural network model of soil behavior. Ellis et al (1995) modeled stress-strain relation of sands using sequential and regular neural network. It is observed that good agreements exist between laboratory data and modeling results. It is also noted that a prescribed strain rate (0.0405 %) has to be defined in order to make prediction with their mode, and pre-identification of different loading stages is required for modeling loading-unloading-reloading characteristics of the sand. These two conditions limit their model only being applicable where the strain rate is 0.0405%, and loading/and unloading stage is known in priori (Parlos et al. 1994). The objective of this paper is to present a recurrent neural network (RNN) model for cohesionless soils. A comparison between the proposed RNN model and a traditional constitutive model – three invariant-dependent cap model was made. A dynamic gradient descent algorithm is used to train the network. It is found that the procedure for implementing the RNN model is much simpler than that for the traditional constitutive model. Only seven basic inputs that can be directly taken from the experimental data are employed in developing the RNN model, and there is no need to find material parameters that are vital to the traditional constitutive models. The implementation of the RNN modeling merely involves repeated computations on several sets of experimental data with a computer. Modeling results show that the characteristics in terms of stress-strain curves of the soil are simulated very accurately by the RNN model. The volumetric changes including compressive and dilative responses during shearing at different stress levels are well represented. Furthermore, the RNN model shows its unique efficiency in capturing rapid variations in volumetric strains during a given unloading and reloading process. In addition, the RNN model overcomes the limitations of fixed strain rate and identification of the various loading stages beforehand for the prediction of the soil behavior subject to different loading conditions.

2 THREE INVARIANT- DEPENDENT CAP MODEL

For the purpose of comparison, we cite here briefly one of the traditional constitutive models, three invariant-dependent cap model proposed by Faruque, et al. (1987, in press). The original cap model developed by DiMaggio et al. (1971) is a two-invariant dependent elasto-plastic model. In the original cap model, the cross-section of the failure envelope on the octahedral plane is always circular that is incidentally not true for all soils, especially for sand. Experimental evidence, in general, suggests that the shape of the failure surface on the octahedral plane changes with the intensity of the compressive mean pressure. At low mean pressure, the surface plots triangular with rounded corners. The failure surface becomes increasingly rounded with the increase in compressive mean pressure. It becomes almost circular at very high compressive mean pressure (Zaman et al. 1996). These features of the failure surface can be simulated by expressing the failure surface in terms of the three stress invariants (Faruque & Zaman, in press; Faruque 1987). In the three-invariant dependent cap model, the failure surface F_f is expressed as:

$$F_f(I_1, J_{2d}, \theta) = g(\theta, I_1) J_{2D} + Ce^{-BI_1} - MI_1 - A \tag{1}$$

where, θ represents a third invariant, defined by:

$$\cos 3\theta = \frac{3\sqrt{3}}{2} \frac{J_{3D}}{J_{2D}^{3/2}}$$

I_1 = the first invariant of stress tensor, $I_1 = \sigma_{ii}$;

$$J_{2D} = (S_{ij}S_{ij})/2, \qquad J_{3D} = (S_{ij}S_{jk}S_{ki})/3 ;$$
$$S_{ij} = \sigma_{ij} - (\sigma_{kk}\delta ij)/3$$

$$g(\theta, I_1) = \frac{2}{\sqrt{3}} \cos\left[\frac{1}{3}\cos^{-1}(-A_t \cos 3\theta)\right]$$

$$A_t = A_0 \exp(-\gamma I_1 / p_a)$$

J_{2D}, J_{3D} = the second and the third invariants of deviatoric stress tensor, respectively;

where A_t, A_o, B,C and M, γ = material constants;

S_{ij} = deviator stress tensor;

P_a = atmosphere pressure.

The equation describing yield cap F_c is given by:

41

$$F_{c0}(I_1, J_{2D}, \theta, k) = g(\theta, L)\sqrt{J_{2D}} * R - \sqrt{(X - L)^2 - (I_1 - L)^2} = 0 \qquad (2)$$

where L is a value of I_1 at the intersection of the cap with the failure surface, R and X are hardening parameters, and X can be further expressed as:

$$X = -\frac{1}{D}\ln(1 - \frac{\varepsilon^{vp}}{W}) + Z$$

where D, W and Z are the material constants describing plastic hardening of the soil, εvp is the plastic volumetric strain.

In order to solve the above equations, one need to incorporate these equations into a formula describing relationship of incremental stress dσ and incremental strain dε, which is given by:

$$\{d\sigma\} = [D^{ep}]\{d\varepsilon\}$$

where [D^{ep}] is a well-known incremental elasto-plastic stress-strain matrix that can be found in any standard reference on elastic constitutive modeling (Desai 1990). In order to use this equation, one has to determine the Young's modulus E and the Poisson's ratio ν beforehand.

It can be seen that there are in all eleven parameters needed to be found for the application of the three-invariant cap mode. These parameters have to be evaluated by conducting some special experiments and optimization techniques (Azeemuddin 1988; Zaman et al. 1982). Specifically, conventional triaxial compression tests can be used to find the Young's modulus E and the Poisson's ratio ν. Failure surface parameters Ff can be solved by running both triaxial compression tests and extension tests, while to find the hardening parameters a series of hydrostatic compression tests should be conducted (Azeemuddin 1988; Faruque 1987). The determination of the parameter values involves a series of tedious tasks, which includes identifying yield points, elastic and plastic strains, determining an average slope of the unloading-reloading portion from the stress-strain curves, etc.

3   RECURRENT NEURAL NETWORK PARADIGM

Neural networks have recently emerged as a very promising tool for various engineering applications. These include pattern recognition, classification, speech recognition, manufacturing process control and material behavior modeling as well (Ghaboussi 1992; Elman 1990; Parlos et al. 1994; Connor et al. 1994; Giles et al. 1994). It is now recognized from the literature that recurrent neural networks (RNN) are often superior to other kinds of networks in modeling nonlinear problems. A typical architecture of the RNN is shown in Figure 1. The initial input **X** is propagated to the hidden layer **Z** through the weights V and bias term V$_o$. The results from the hidden layer are then transmitted to the output layer **Y** through weights W and bias W$_o$, and feedback to the input layer as external data inputting to the hidden layer again. The algorithm of the RNN shown in Figure 1 is given by:

$$Z_{jt} = F_1(\Sigma V_{ijt} * X_{it} + V_{0t} + \Sigma V_{Zt\text{-}i(t-1)} * Z_{i(t-1)}) \qquad (4)$$

$$Y_{kt} = F_2(\Sigma W_{jkt} * Z_{jt} + W_{0t})$$
$$= F_2(\Sigma W_{jkt} * F_1(\Sigma V_{ijt} * X_{it} + V_{0t} + \Sigma V_{Zt\text{-}i(t-1)} * Z_{i(t-1)}) + W_{0t}) \qquad (5)$$

where

F_1, F_2 = the activation functions from the input layer to the hidden layer and from the hidden layer to the output layer, respectively;

V_{ijt} = the weight connecting node j in the hidden layer with the node i in the input layer at t^{th} epoch;

W_{jkt} = the weight connecting node k in the output layer with the node j in the hidden layer at t^{th} epoch;

V_{ot} , W_{ot} = the biases at t^{th} epoch in the hidden layer and output layer, respectively.

$\overset{\cdot}{V}_{zt\text{-}i(t\text{-}1)}$ = the weight between output from hidden layer at $(t\text{-}1)^{th}$ epochs and hidden layer net at t^{th} epochs.

The updates of the weights can be expressed as follows:

$$V_{ijt} = V_{ijt\text{-}1} + \Delta V_{ij} ; \tag{6}$$

$$W_{jkt} = W_{jkt\text{-}1} + \Delta W_{jk} ; \tag{7}$$

$$V_{ot} = V_{ot\text{-}1} + \Delta V_o; \tag{8}$$

$$W_{ot} = W_{ot\text{-}1} + \Delta W_o; \tag{9}$$

The weight changes ΔV_{ij}, ΔW_{jk} can be derived by minimizing the sum of squared errors between target value T and output value Y using delta rules (Vogl *et al.*1988; Widrow & Lehr 1988) taking into consideration of a momentum parameter. Specifically, updates of the weights can be written as follows:

$$\Delta V_{ij} = -\alpha * \frac{\partial E}{\partial V_{IJ}} + \mu * \Delta V_{ij(t\text{-}1)}$$
$$= \alpha * \Sigma (T_k - Y_k) * F_2'(Y_k) * W_{jk} * F_1'(Z_j) * X_i + \mu * \Delta V_{ij(t\text{-}1)} \tag{10}$$

$$\Delta W_{ij} = -\alpha * \frac{\partial E}{\partial V_{jk}} + \mu * \Delta W_{ij(t\text{-}1)}$$
$$= \alpha * (T_k - Y_k) * Z_j * W_{jk} * F_2'(Y_k) + \mu * \Delta W_{ij(t\text{-}1)} \tag{11}$$

$$\Delta V_0 = \alpha \, \Sigma (T_k - Y_k) \, W_{jk} \, F_1'(Z_{netj}) \tag{12}$$

$$\Delta W_0 = \alpha \, (T_k - Y_k) \, F_2'(Y_{netj}) \tag{13}$$

where

$E = 1/2 * \Sigma_k (T_k - Y_k)^2$ is the sum of squared errors, α being learning rate;

F_1', F_2' = the derivatives of activation function of F_1 and F_2, respectively;

μ = a momentum parameter.

The use of momentum term allows the network to make reasonably large weight adjustments and to proceed not in the direction of the gradient, but in the direction of a combination of the current gradient and the previous direction of weight correction, which is helpful in preventing networks from getting stuck in a local minimum.

DATA BASE

The implementation of RNN modeling is done by using laboratory test data for a Dhahran dune sand that is used extensively in the construction of road base and subbase in the Mid East. The sand, with D_{10} =0.15 mm, a uniformity coefficient of 1.84, and an average relative density of 60%, was tested using a stress controlled cylindrical triaxial device (Azeemuddin, 1988). Eight sets of data from the triaxial compression tests conducted at different confining stresses are employed in the RNN modeling, four of which are used as training data, and the other four sets are used as testing data, as listed in Table 1.

Table 1 Experimental data used for the RNN modeling

Test #	C1*	C2	C3*	C4*	C5	C6	C7	C8*
$\sigma_{30, kPa}$	35	69	104	172	241	276	345	690

*Note: Data with * are used for training process.*

σ_{30} is initial confining pressure.

4 RNN SIMULATION OF THE SOIL BEHAVIOR

It is necessary to identify input and output variables and further design an architecture of the network to activate a RNN model.

(1) <u>Input variables and output results</u>: Appropriate selection of the input and output forms a basis for successful simulation of the RNN model. In stress controlled shearing tests, the stress increment is given and therefore the stress at a specific time is fixed. So, the deviatoric stress $dev\sigma_i$ and major principal stress σ_{1i} and their increments $\Delta dev\sigma_i$ and σ_{1i} should be the input variables. It is worth mentioning that the stress increments are not a constant in the RNN model, because of the laboratory experimental data have been sifted/and reduced substantially for the sake of saving space and computer time. For loading stage, the value of $\Delta dev\sigma_i$ ranges from 27 kPa to about 100 kPa. For unloading stage, the $\Delta dev\sigma_i$ varies between -60 kPa to -50 kPa. Also inputs are previous strains including axial and volumetric strains since strain history will greatly affect soil behavior in the stress controlled tests. The output, from the problem itself, should be current axial strain ε_1 and volumetric strain ε_v. In addition to this, relative density D_{ri} of the soil is considered as an input because it changes during the drained tests. Specifically, the input $\mathbf{X}$ and output $\mathbf{Y}$ shown in Figure 1 are represented as follows:

$$\mathbf{X} = \{dev\sigma_i, \Delta dev\sigma_i, \sigma_{1i}, \Delta\sigma_{1i}, \varepsilon_{1i}, \varepsilon_{vi}, D_{ri}\} \tag{14}$$

$$\mathbf{Y} = \{\varepsilon_{1i+1}, \varepsilon_{vi+1}\} \tag{15}$$

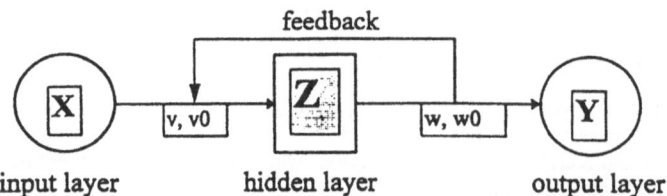

Figure 1 Typical Architechture of Recurrent Neural Network

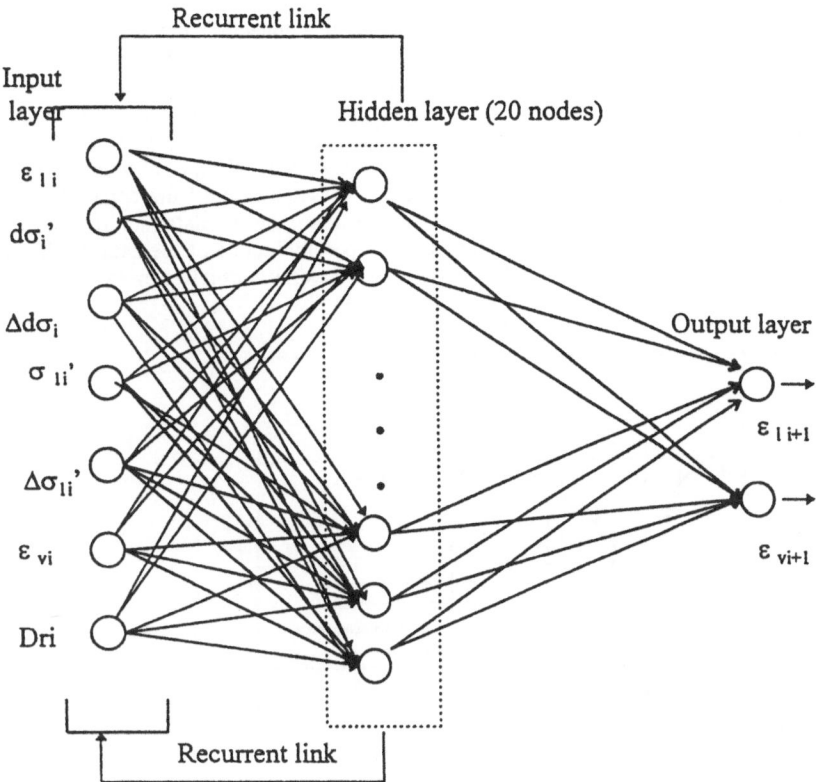

Figure 2 Architecture of 7x20x2 Recurrent Neural Network Model
where:ε_{1i},ε_{vi} =axial strain and volumetric strain, respectively, at i step;
$d\sigma_i{}'$, $\sigma_{1i}{}'$ =deviatoric and major principal stress, respectively, at i step;
$\Delta d\sigma_i$, $\Delta\sigma_{1i}{}'$=increment of $d\sigma_i{}'$, $\sigma_{1i}{}'$ at i step; and Dri =relative density of
the sand at i step loading; ε_{1i+1} and ε_{vi+1}= axial and volumetric strain,
respectively, at i+1 step.

Clearly, in this case the target value T is assumed the experimental values of axial and
volumetrical strain.

(2) <u>Architecture of the model</u>: One hidden layer with connections to the both input layer and
output layer is used in the model because it is considered as very effective in dealing with any
nonlinear problems (Vogl *et al*. 1988), (Widrow & Lehr, 1990). The number of nodes in the
hidden layer is determined by a trial and error method. For a specified epochs (2000 epochs), the
network was trained preliminarily with different number of nodes ranged from 4 to 30 nodes.
The sum squared errors thus obtained are between 1.5 to 3.8 and the least one occurred in a
network with twenty nodes in the hidden layer. So a 7x20x2 network was set up as shown in
Figure 2. It should be noticed that once the training process is initiated, the 7x20x2 network
becomes a 27x20x2 network since the outputs from hidden nodes are fed back to the input layer
as input data again. Therefore, there are 27x20 weights and 20 bias values connecting the input
layer and hidden layer, and 20x2 weights and 2 bias values between the hidden layer and output
layer. Appendix A gives a list of the weight and bias values.

The hyperbolic tangent function ($F_1 =(e^x-e^{-x})/(e^x+e^{-x})$) is chosen as an activation function from
the input layer to the hidden layer, since it results in a fast convergence of the network as

compared with the sigmoid activation function. While in the connection between the hidden layer and the output layer, a linear transform function is adopted for the sake of simplicity.

(3) <u>Training process</u>: The training process is initialized by inputting a set of "comprehensive" data to the network. The so called "comprehensive" data imply that the data are most representative and contain all the necessary information for the given problem. A network trained with the comprehensive data can have strong predictability. The flexibility of the network itself ensures that the network can be trained ceaselessly until the comprehensive data are included in the training data base. As indicated in Table l, the test data numbered as C1, C3, C4 and C8 are used as training data since they contain possible range of the data. Each data set contains about 1100 data. It is necessary, from our experience, to normalize every set of data

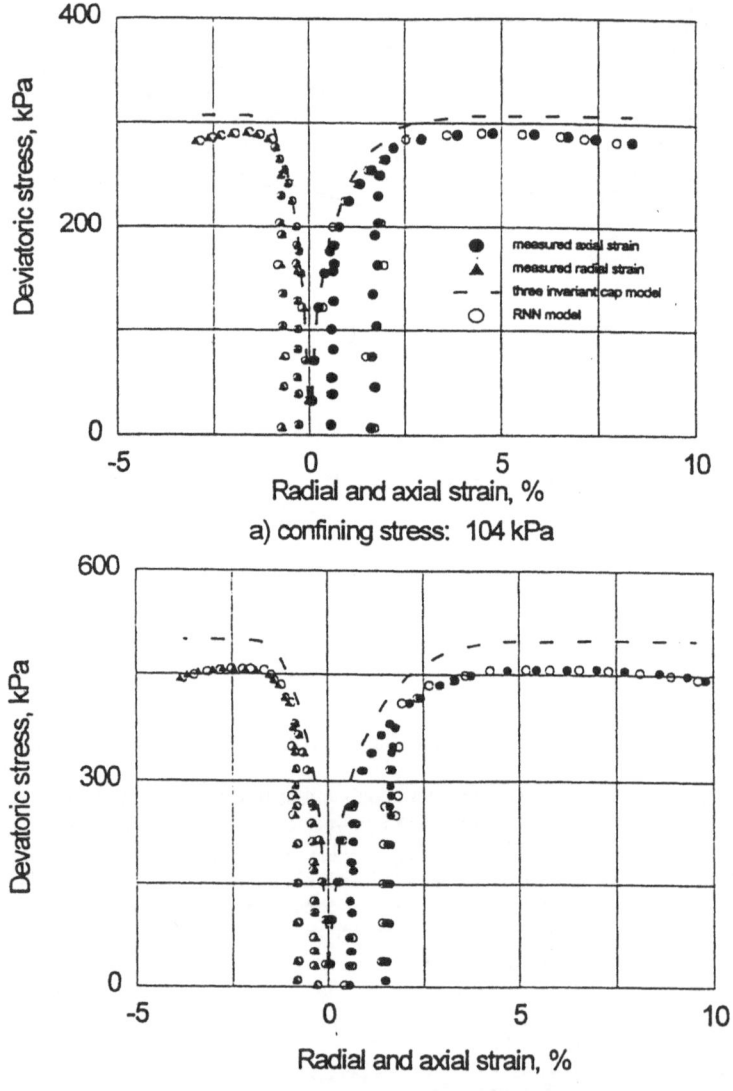

a) confining stress: 104 kPa

b) confining stress: 172 kPa

Figure 3 Simulation of stress-strain relationships of experimental data with the RNN model together with the three invariant-based cap model

46

with respect to its maximum and minimum values before initiation of training process. After normalization, each data value is presented within a range of (0, 1), with their maximum and minimum values equaling 1 and 0, respectively. This preprocessing of data guarantees that the network operates in a more efficient and more reliable manner. The training process was performed with Matlab on a PC/Gateway2000 computer. In order to enhance flexibility of the network, the training process of the network was also designed' as a dynamic system that can make use 12 of previous knowledge in the case of training new data and when more epochs of training are required. It was found that when training reached 6,000 epochs the sum squared error reduced 0.05, which took about 180 minutes. After 6,000 epochs, the sum squared error was almost unchanged with the increase of the epochs.

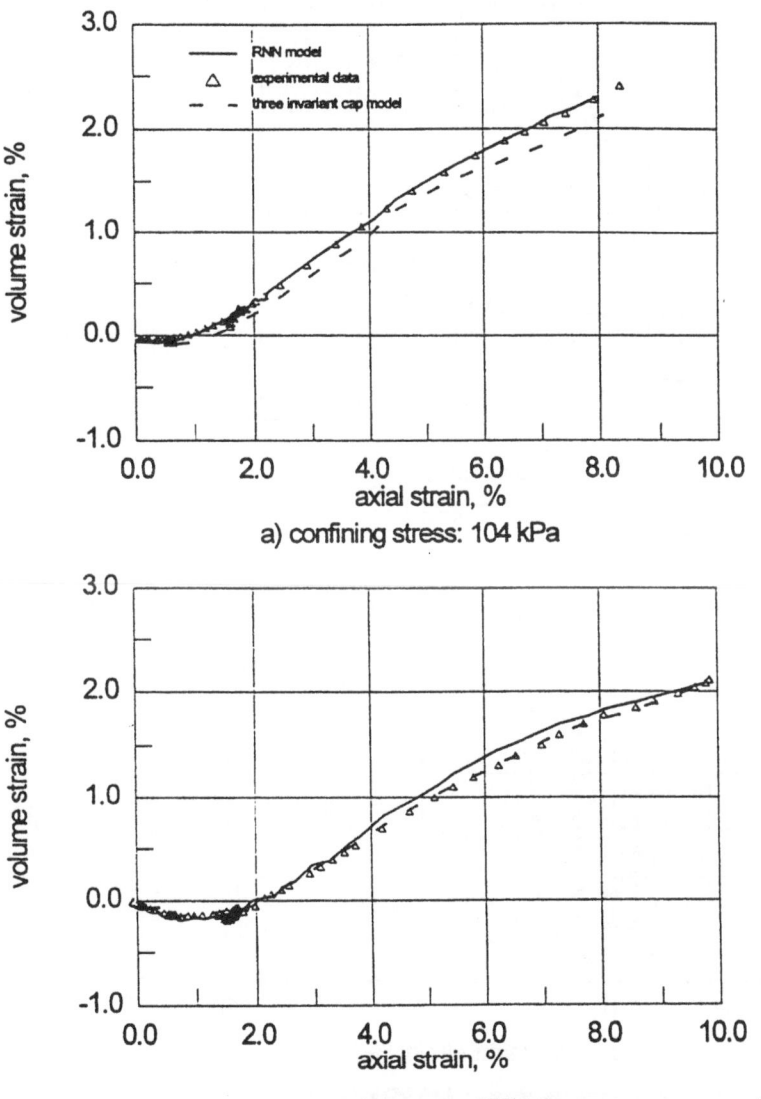

a) confining stress: 104 kPa

b) confining stress: 172 kPa

Figure 4 Simulation of volumetric strain vs. axial strain of experimental data with the RNN model together with the three invariant cap model

(4) <u>Testing of networks</u>: After completion of training process, a set of weight and bias values are produced. These values connecting different layers in the network are used for testing of the designed network. The testing process contains simulation of network with the trained data and prediction of network with testing or untrained data. During the testing process, there still exist chances to update the weights and biases of the network. In general, more training epochs with the original training data are required if there is non-eligible deviation from the true data in the simulation process. However, if one is not satisfied with the prediction results for a specific set of data, then this set of data may work as training data to train the network again to produce new connections which are more knowledgeable than the old one.

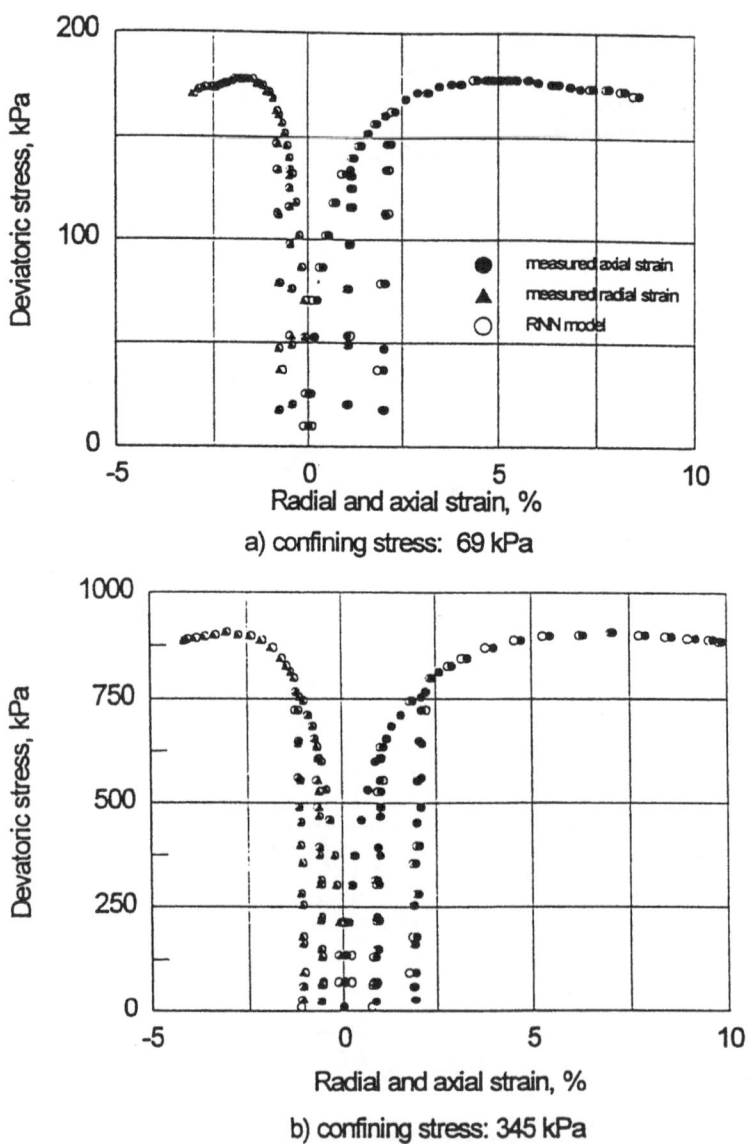

a) confining stress: 69 kPa

b) confining stress: 345 kPa

Figure 5   Prediction of stress-strain relationships of using untrained experimental data with the RNN model

48

5   MODELING RESULTS

The RNN modeling gives simulation and prediction results with trained and testing data, respectively. For the purpose of comparison, two sets of results obtained from the three invariant dependent cap model and the RNN simulation are plotted against the experimental data.

Simulation with trained data

Figure 3 shows typical RNN simulation results in terms of stress-strain behavior of the dune sand with the trained experimental data at confining stress of 104 kPa and 172 kPa. The radial

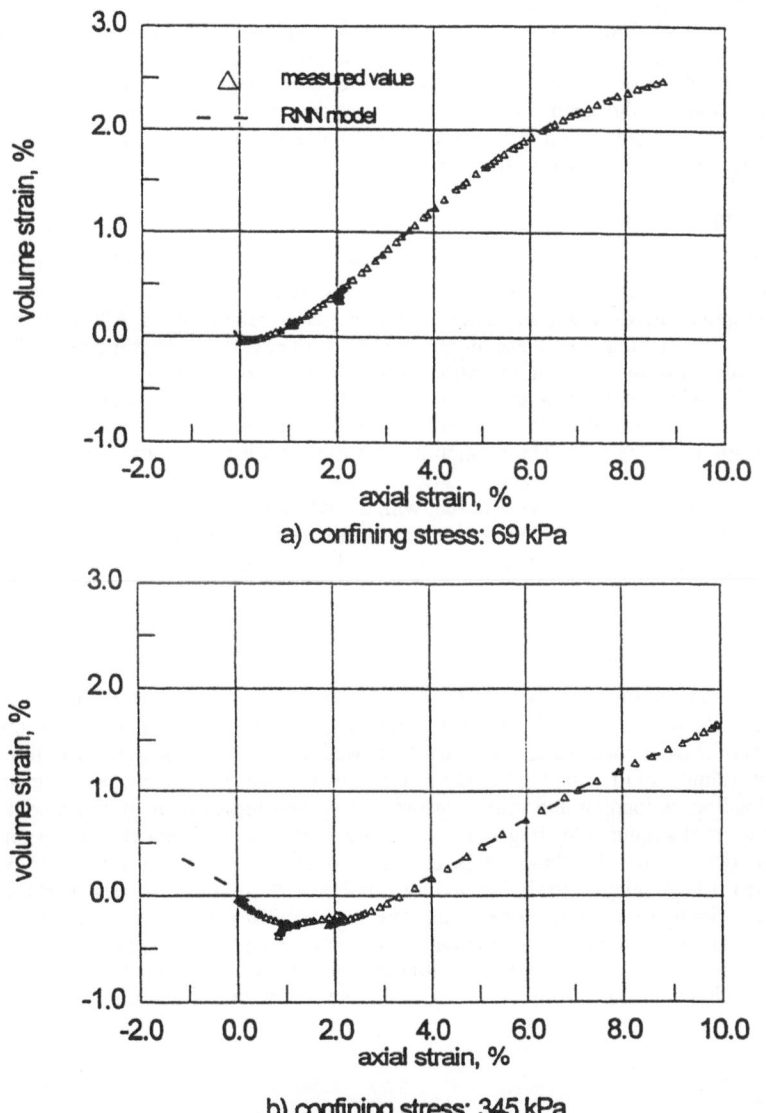

Figure 6 Prediction of volumetric strain vs. axial strain of using untrained experimental data with the RNN model

49

strain (ε_r) shown in the figure is calculated from volumetric and axial strain by using a common formula: $\varepsilon_r = (\varepsilon_v - \varepsilon_v)/2$. From Figure 3 one can see that the RNN simulations of stress-strain behavior are almost identical to the measured values in any course of loading, unloading and reloading. The various stress-strain behavior including linear elastic, plastic, hardening and softening are all represented well by the RNN model. This agreement demonstrates that the RNN model has a strong capability to simulate complex stress-strain behaviors of the soil. The dashed lines shown in the figure are obtained from the three invariant-dependent cap model. Note that only continuous loading curves were modeled by the three invariant-dependent cap model because of extreme difficulties in modeling unloading and reloading situations for this model. Figure 3 also shows that the three invariant-dependent cap model is unable to simulate softening behavior of the soil. It is evident that the RNN model produces much more accurate results than the traditional constitutive model.

Figure 4 shows the typical simulation results of volumetric response of the soil tested at the same confining stresses as shown in Figure 3. From the figure one can see that both dilative and compressive volume changes are simulated by the RNN model with minor deviation in Figure 4 b. Also an exciting feature is the simulation of rapid variations in volumetric strains during unloading-reloading process. This observation is of particular significance in understanding mechanism of dynamic loading like seismic and vehicular traffic loading.

Prediction with testing data

Figures 5 and 6 demonstrate the predictive capability of the RNN model with untrained experimental data at confining stress of 69 kPa and 345 kPa. Figure 5 presents two sets of stress-strain curves, while Figure 6 gives two sets of axial-volumetric strain relationships. Although the RNN model did not learn anything from these untrained data, the network could still predict the soil behavior very precisely based on the information learned from the training data. As did with the training data, both loading, unloading and reloading features are represented with sufficient accuracy in the prediction curves, and both dilation and compression volume changes are well predicted with the testing data as well. The prediction with other testing data produces similar accurate results.

From the results presented above, it is evident that with the RNN modeling one does not need to find a series of parameters from which mathematical equations are proposed. This is one of the important advantages of the RNN modeling over the traditional constitutive modeling.

6   CONCLUSIONS

The framework of recurrent neural network (RNN) modeling for cohesionless soils is presented. A 7x20x2 recurrent neural network is designed for the modeling of a cohesionless soil, Dhahran dune sand. Eight sets of triaxial compression test data were used in the RNN modeling, with four sets of data used for training and testing the network, respectively. The stress-strain behavior of the soil including loading, unloading and reloading process is well represented by the model. There is not necessary to distinguish loading stage from unloading and reloading conditions for simulationland prediction of the soil behavior subject to different loading conditions in the RNN model. The predicted 15 volumetric strains including dilative and compressive strains show excellent consistency with those of the measured data. The variations in volumetric strain during unloading and reloading stages are precisely captured by the RNN model. These results illustrate that the RNN model is efficient in simulating nonlinear behavior of a cohesionless soil. As compared with the traditional constitutive modeling, the RNN modeling is much simpler and much more efficient.

ACKNOWLEDGMENTS

The authors would like to express their thanks to Dr. Mohammed Azeemuddin for providing the experimental data used in this paper. Financial support from the Oklahoma Department of

Transportation and the School of Civil Engineering and Environmental Science at the University of Oklahoma, Norman is also gratefully acknowledged.

REFERENCES

Abduljauward, S. N., Faruque, M. O. & Azeemuddin, M. & Al-Ghamedy, H. N. Three- Invariant Dependent Cap Model with Application to a Cohesionless Soil. Australian Civil Engineering Transactions, TEAust, Vol. CE34 No. 4,1992, pp. 335-367.

Azeemuddin, M. Constitutive Modeling of Dhahran Dune Sand Using Cap Model. M.S. Thesis, Univ. Of Pet. & Min., Dhahran, Saudi Arabia, 1988.

Connor, J. T., Martin, R. D. & Atlas, L. E. Recurrent Neural Networks & Time Series Prediction. IEEE Transactions on Neural Network, Vol. 5, No. 2, 1994, pp. 240- 254.

Desai, C. S. Modeling and testing : Implementation of numerical models and their application in practice. Numerical Method and Constitutive Modeling in Geomechanics. Edited by Desai and Gioda, Springer-Verlag, Wien-New York, 1990, pp.1-168.

Desai, C. S. & Watagala, G. W. Constitutive model for cyclic behavior of Clays. J. of Geotech. Eng. Div., ASCE, Vol.119, No. 4,1993, pp.714-729.

DiMaggio, F.L. & Sandler, I. S. Material Model for Granular Soils. J. Eng. Mech. Div., ASCE, Vol. 97, No. EM3,1971, pp.935-950.

Drucker, D. C., Gibson, R. E. & Henkel, D. J. Soil Mechanics and Work Theories of Plasticity, Proc. ASCE, Vol.122,1957, pp.338-346.

Duncan, J. M. & Chang, C. Y. Nonlinear Analysis of stress and strain in soils. J. Soil Mechanics and Foundations Division, ASCE, Vol. 96, No. 5M5, 1970, pp.1629-1653.

Ellis, G. W., Yao, C. Zhao, R. and Penumadu, D. 1995. Stress Strain Modeling of Sands Using Artificial Neural Networks. Journal of Geotechnical Engineering *Division*, ASCE Vol. 121, No 5 pp. 429-435.

Faruque, M. O. A Third Invariant Dependent Cap Model for Geological Materials. Soils and Foundations, Vol. 27,1987, pp.12-20.

Faruque, M. O. & Zaman, M. M. Modeling Stress Strain and Dilatant Behavior Cohesionless Soil. Computers and Geotechnics, in press. 16

Fukushima, S. & Tatsuoka; F. Strength and Deformation Characteristics of Saturated Sand at Extremely Low Pressures. Soils and Foundations, Vol. 24, No. 4, 1984, pp. 30-48.

Ghaboussi, J. Neural-Biological Computational Models with Learning Capabilities and Their Applications in Geomechanical Modeling. Proc. of Recent and Future Trends in Geomechanics in the 21st Century, U.S. Canada Edited by Zaman, Desai and Selvadurai, Norman, Oklahoma,1992, pp.131-134.

Ghaboussi, J. Potential application of neural -biological computational models in geotechnical engineering. Proc. 4lth Int. Sym. On Neumerical Models in Geomechnics, Swansea, U.K.1992.

Ghaboussi, J., Garrett, J. H. & Wu, X. Knowledge-based Models of Material Behavior with Neural Networks. J. Of Eng. Mech. Div., ASCE, Vol. 117, No.l 1991, pp.132-153.

Giles, C. L., Kuhn, G. M. & Williams, R. J. Dynamic Recurrent Neural Networks: Theory and Applications. IEEE Transactions on Neural Network, Vol. 5, No. 2, 1994, pp.153-160.

Lade, P. V. Elastic-Plastic Stress-Strain Theory for Cohesionless Soil with Curved Yield Surfaces. Int. J. of Solids and Structures, Vol.13,1977, pp.1019-1035.

Loung, M. P. Stress-Strain Aspects of Cohesionless Soils Under Cyclic and Transient Loading. Proc. Int. Sym. On Soils Under Cyclic and Transient Loading, G. N. Pande and O. C. Zienkiewicz (Eds.). Vol. l, A. A. Balkema, Rotterdam,1980, pp. 315-324.

Parlos, A. G., Chong, K. T. & Atiya, A. F. Application of the Recurrent Multilayer Perceptron in Modeling Complex Precess Dynamics. IEEE Transactions on Neural Network, Vol. 5, No. 2,1994, pp. 255-285.

Rumelhart, D., Hinton, G. E. & Williams, R. J. Learning internal representations error propagation. Parallel Data Processing, Rumelhart and McClelland eds., M.I.T. Press, Cambridge,1986, pp. 318-362.

Vogl, T. P., Mangis, J.K., Rigler, A. K., Zink, W.T. & Alkon, D. L. Accelerating the convergence of the back propagation method. Biological Cybernetics, Vol. 59, 1988, pp. 257-263.

Widrow, B. & Lehr, M. A. 30 Years of Adaptive Neural Networks: Perceptron Madaline and Backpropagartion. Proc. IEEE., 78.(9),1990, pp.1415-1442.

Zaman, M. M., Desai, C. S. & Faruque, M. O. An Algorithm for Determining Parameters for Cap Model from Raw laboratory Test Data. Proc. 4th Int. Conf Numerical Methods in Geomech., Edmonton, Canada,1982.

Zaman, M., Faruque, M. O. & Abdulraheem, A. Three Invariant Characteristic State Model For Cohesionless Soil, in review, Int. J. Num. and Anal. Met. in Geomech.,1996.

Appendix A Weights and biases of the RNN model developed in this study

Weights connecting input layer and hidden layer (V27x20)=

-0.3160	-0.1830	-0.3350	0.2920	-0.2460	-0.0704	-0.3220	0.1110	-0.1540	-0.2780	-0.3550	-0.1440	0.0413	0.3820	0.3360	0.0943	-0.6420	-0.0890	-0.6360	0.2570
0.1020	0.3540	-0.0237	-0.1750	0.1120	-0.1510	-0.1030	0.4910	0.4220	0.3560	0.2380	-0.2420	-0.3170	0.4450	0.0731	0.0170	0.4100	0.2150	0.3140	0.0422
0.0844	-0.4340	0.4080	0.2360	-0.2850	-0.4550	-0.1000	-0.3510	-0.3420	-0.0743	-0.4570	0.1370	0.3190	-0.0170	0.2070	0.1920	0.3150	-0.5180	-0.3270	0.1550
-0.1670	-0.4120	-0.2280	0.4960	-0.2430	-0.2060	0.0885	-0.3590	-0.4660	0.1880	0.1250	0.3080	-0.3990	0.0479	0.3590	0.5730	-0.1260	-0.0143	0.3330	0.4010
0.1160	0.1380	-0.0520	-0.0856	0.5550	-0.2780	0.1410	-0.1820	0.0693	0.1160	0.2380	-0.4970	-0.0350	0.0743	0.2000	-0.0994	0.1800	0.1820	-0.0261	-0.3390
0.0034	0.2400	-0.4020	-0.4090	-0.2450	0.0284	-0.4600	-0.3000	-0.3420	-0.1890	-0.0253	0.1130	0.1670	0.3480	0.2580	0.0582	0.2650	0.1640	0.2120	-0.0994
-0.0124	0.0179	0.1890	0.0849	-0.2070	0.1250	-0.2840	0.1460	-0.0062	0.5650	-0.3150	-0.1550	0.1210	0.4070	0.3210	0.2720	-0.2440	0.0143	-0.3300	-0.3240
0.2200	-0.1040	0.1080	-0.2070	0.2940	-0.1390	0.1800	-0.1930	0.2140	-0.0138	-0.2690	-0.1010	0.1220	-0.0280	-0.2440	0.1750	-0.1590	-0.0226	-0.1060	-0.0934
0.1560	-0.1320	-0.0521	0.0182	-0.0207	-0.0936	0.1320	-0.1510	0.2440	0.1870	-0.0742	-0.0743	0.2090	-0.2180	-0.1580	-0.2680	-0.1190	-0.0041	-0.0488	0.1620
-0.0481	-0.0008	-0.0193	-0.1070	-0.1760	0.1320	0.1410	0.1410	0.1360	0.1160	-0.1100	0.0368	-0.3750	0.2230	0.0164	0.1460	-0.0916	0.2900	0.0364	-0.0211
0.1670	0.0223	0.2090	0.1970	0.0741	0.1960	0.1450	0.0449	0.0921	-0.1490	-0.1310	0.1320	-0.1650	0.2020	-0.2920	-0.1330	-0.1890	-0.1700	-0.0699	-0.0445
0.0909	0.1150	-0.0510	0.1500	0.2090	-0.0391	0.2050	-0.0988	-0.0930	-0.1320	0.1270	-0.1490	0.1320	-0.0257	0.1460	-0.1300	-0.2040	0.1010	-0.2460	0.2290
0.2630	0.0673	0.2050	-0.1850	0.2460	-0.1850	0.1630	0.0254	0.0955	0.2200	-0.2960	-0.0552	0.0594	0.1460	0.0070	-0.0607	-0.2030	0.1070	0.1180	0.1150
-0.0606	-0.0790	0.1630	0.0254	0.2130	-0.1320	0.0836	-0.0098	0.0622	0.1300	-0.1180	-0.0324	-0.0090	-0.1190	0.0259	-0.0346	-0.0926	0.1440	0.0846	0.1740
-0.0601	0.1330	0.0836	-0.0098	0.1290	-0.0930	-0.1990	0.2100	-0.0719	-0.2940	-0.2940	-0.0738	-0.1190	0.0070	-0.0346	0.1690	-0.2420	0.2350	-0.1080	-0.2210
-0.0989	-0.1520	-0.1990	0.1040	0.0661	-0.1350	-0.0756	0.1040	-0.1350	-0.2630	-0.2760	-0.2940	0.0977	-0.0869	0.1690	0.1700	-0.0346	0.1440	-0.1080	-0.1460
0.1650	0.0189	0.0661	0.0275	0.1040	0.2100	0.2100	0.0272	0.0041	0.0160	-0.1510	-0.2760	-0.1100	0.0605	0.1700	0.1690	0.1430	-0.0115	0.0947	0.2140
-0.0730	0.2130	-0.0599	0.0507	0.1040	0.1040	0.1040	0.2390	0.0366	0.0272	0.0160	-0.2630	-0.1670	-0.2130	0.0489	0.1700	-0.0323	0.2420	-0.0423	0.2040
-0.2720	0.2220	-0.0402	-0.1080	-0.1910	0.0272	0.2390	0.1040	0.1280	0.2100	0.0041	0.2870	-0.1400	0.0957	0.0202	0.0489	-0.1620	0.2250	-0.1460	-0.1800
-0.1130	-0.2150	-0.1910	0.1410	-0.2150	0.0928	0.1050	0.1050	0.1230	0.1590	0.1590	0.0414	-0.1370	0.0202	-0.2070	-0.1880	-0.1620	-0.0331	-0.1460	-0.1800
0.1800	0.0351	0.1570	0.2560	0.1130	-0.0288	-0.0712	0.0696	0.0638	0.0535	0.2270	-0.1370	0.2400	-0.2070	-0.1370	0.1820	-0.2200	0.2250	0.0819	0.0348
-0.2390	0.1360	0.0174	-0.0788	-0.0538	0.2240	0.2240	0.2450	-0.1630	0.0806	0.0647	-0.0266	0.0250	-0.0980	0.1820	0.1140	0.0693	-0.0605	-0.1610	0.2120
-0.1390	0.1650	0.0066	-0.2480	0.0066	0.2130	0.2130	0.1880	-0.2010	0.2450	0.2290	0.0493	-0.2400	-0.1400	0.1140	0.0693	0.1720	-0.0117	-0.0117	0.0255
-0.1910	0.1480	0.1270	0.1880	0.1270	-0.1850	-0.0860	0.2190	-0.2580	-0.1850	0.0187	0.1270	-0.1730	-0.2310	-0.1450	0.1860	-0.1060	0.1500	0.1390	0.0585
0.0428	-0.0713	-0.1990	-0.0666	-0.1990	0.2190	0.2190	0.0772	-0.1250	0.1840	-0.0923	-0.1430	0.1720	0.1470	0.1860	-0.1450	-0.1180	-0.1280	-0.1950	0.0258
-0.2220	0.2260	0.2210	-0.1440	0.2210	0.0772	0.1690	0.1690	-0.1800	0.1500	0.1500	-0.0442	-0.0078	0.1210	-0.0746	-0.1860	-0.1920	-0.1920	-0.0729	-0.1590
0.2330	0.1340	0.1070	-0.0732	0.1070	0.1690	-0.1510	0.2010	-0.2710	0.0995	-0.2710	0.1890	0.1400	0.1010	-0.0815	-0.0433	-0.1990	-0.1990	-0.1170	-0.2350

Biases in the hidden layer (b20x1)' =

0.7489	0.6569	0.3006	0.3246	1.0841	0.6835	-0.0801	-0.1908	1.0929	-0.5810	-0.2595	-0.4604	0.2824	-0.3726	-0.6296	-0.3605	0.1465	0.1636	0.0233	-0.2687

Weights connecting hidden layer and output layer (W20x2)' =

-0.2470	0.8080	0.3560	0.6070	-0.3530	0.2870	-0.7680	-0.6010	-0.6620	0.4200	-0.0767	-0.5390	0.2920	0.7610	-0.6710	0.8290	-0.9350	-0.6120	-0.4850	0.2340
-0.4550	0.1900	-0.5650	-0.4250	0.2110	0.6130	-0.2530	-0.4410	-0.6030	-0.8110	0.8490	-0.5600	-0.8630	-0.6560	0.6410	-0.8210	0.6330	0.5600	-0.2800	0.2740

Biases in the output layer (b2x1)' =

-0.7863 0.9193

Artificial Intelligence and Mathematical Methods in Pavement and Geomechanical Systems, Attoh-Okine (ed.)
© *1998 Balkema, Rotterdam, ISBN 90 5809 028 0*

On the use of BPNN in liquefaction potential assessment tasks

Yacoub M. Najjar & Hossam E. Ali
Department of Civil Engineering, Kansas State University, Manhattan, Kans., USA

ABSTRACT: Back-Propagation NeuroNet (BPNN) approach was used herein to characterize the soil liquefaction potential using 61 field data sets representing various earthquake sites from around the world. Various combinations of input parameters were used to model the soil liquefaction potential in order to investigate the impact of each variable on the liquefaction potential. For brevity, only the most promising liquefaction prediction models are presented in this paper. The accuracy of the designed BPNNs was validated, when possible, against an additional 44 records never used before in training or testing stages. Furthermore, the prediction accuracy of the BPNN-based models was compared with predictions obtained using other approaches.

1 BACKGROUND

Liquefaction is considered as one of the most severe seismic hazards resulting from earthquakes. Recognizing that soil liquefaction can lead to catastrophic damage to all types of structures and may significantly contributed to the loss of human life; concentrated efforts statistical, analytical, probabilistic, or numerical approaches have been conducted by various researches in order to establish useful criteria that can aid in distinguishing between potential liquefaction and nonliquefaction situations based on given earthquake, site and soil-related information.

First attempts to assess soil liquefaction potential were based on empirical correlations (rules) deduced from field observations of previous earthquakes. Due to the unreliability of these empirical methods, many semi-theoretical approaches were developed in the developed by Seed & Idriss (1971). This approach was based on the cyclic stress analysis. This approach was later modified to utilized the results of in-situ tests such as standard penetration test (SPT) or static cone penetration test (CPT) instead of laboratory tests [Christian & Swiger (1975), Seed *et al.* (1983), Seed *et al.* (1985), Stark & Olson (1995) and Trifunac (1995)]. These methods incorporated the cyclic horizontal stresses with statistical analyses to develop criteria that can discriminate between liquefaction and non-liquefaction possibilities. Another method, based on the energy dissipation principle, was developed by Berrill & Davis (1985). In this method, the authors used the hypothesis that the dissipated seismic energy and the increase in pore water pressure are proportional in order to establish a new criterion for assessing liquefaction potential. Another approach combining discriminant-regression analysis and probabilistic and numerical techniques have been utilized/developed to assesss liquefaction potential. Among these are the fuzzy logic approach [Elton *et al.* (1995), El Zahaby & Rahman (1996)] and the Black-Propagation NeuroNet (BPNN) technique [Goh (1994), Agrawal *et al.* (1997), Najjar and Ali (1998), Ali and Najjar (1998)].

This paper summarizes the research experience of the Kansas State University research team in using BPNN approach to assess the soil liquefaction potential based on SPT measurements.

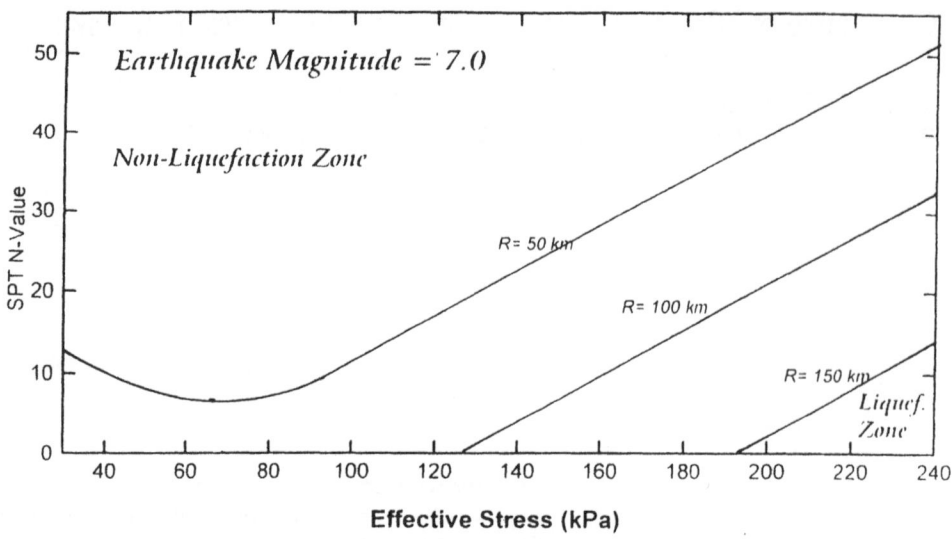

Figure 1 : Liquefaction potential assessment chart based on BPNN-model A for earthquake magnitude = 7

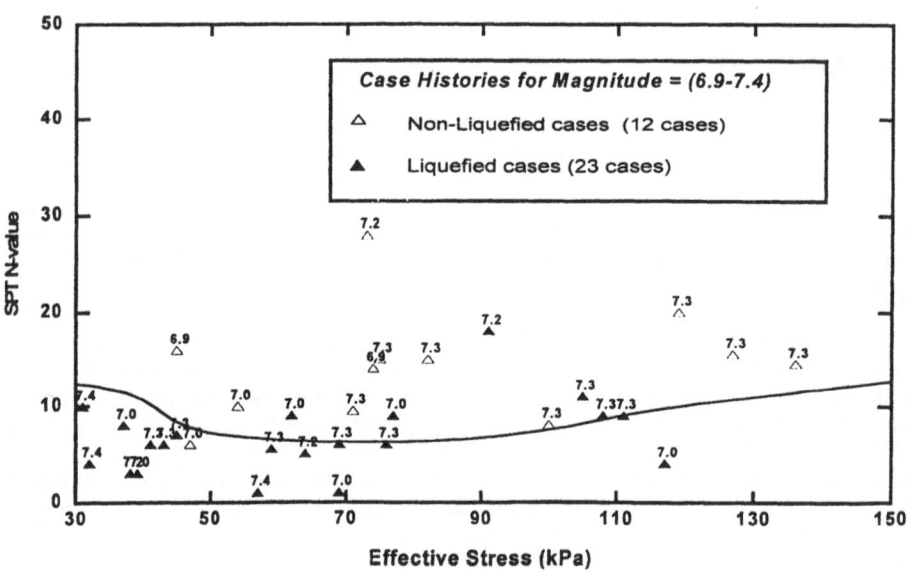

Figure 2 : Overlay of actual case histories on liquefaction potential discriminant chart of BPNN-model B for earthquake magnitude = 7

The three most promising stand-alone BPNN-based models along with a hybrid BPNN-based model are presented in this paper. Results obtained from this study, comparison with assessments obtained from other methods related to this comparison are also presented.

2   BPNN-BASED MODEL DEVELOPMENT

2.1   BPNN Approach

BPNN are massively parallel computational techniques capable of mapping and capturing many features and sub-features (the pattern) embedded within a large set of data that yield certain output. A network that has successfully captured the governing relationships between the input and the output can be used as a prediction tool for cases when the output solution is not available. The structure of BPNN consists of an input layer made of a number of input nodes that are presumed (by the designer) to account for and explain the variability observed in the output(s) of a specific problem. An output layer is designed to contain output nodes (variables) that are pursued for the problem at hand. An intermediate layer(s) (i.e. the hidden layer(s)) contains a number of units that have no interaction with the external environment but are interconnected with the nodes of other layers. The nodes in a certain layer are connected to all nodes in the neighboring layers. The computational efficiency of the BPNN lies in its interconnecting nodes. Further details on BPNN topology, training, and testing can be found elsewhere [e.g., Hassoun (1995); Simpson (1990)]. Among the several neuronets that have been developed is the three-layered feed-forward errorbackpropagation network with supervised training. This technique has been successfully used in modeling various applications in civil engineering [Najjar & Basheer (1996), Najjar *et al.* (1996), Ellis, *et al* (1995)].

2.2   Modeling

The complete set of data (a total of 105 records) from previous case histories reported by El Zahaby (9) was utilized in this study. These records were divided into three groups. The first group consisted of 44 records (≈70% of the 61 available complete data sets) which were used for network training purpose. The second group representing about 30% of the available complete sets (17 records) was used for testing the prediction accuracy of the developed neural networks. The final group consisted of additional 44 records, which was used to validate the performance of the developed networks. Due to the fact that the third group had some missing input variables, the validation process was not applicable to all neuronet-based models.

The output layer, in all developed ANN models, had only one node which represents the liquefaction potential. During the training stage, the output of this node was restricted to two discrete values (i.e., 0 for non-liquefaction cases and 1 for liquefaction cases). However, during the testing and validation stages, the output value was allowed to take a continuos range from 0 to 1. The value of 0.5 was chosen in this case to be the threshold between the liquefaction and non-liquefaction potential zones.

Several models, in this study, have been developed using a combination of different in put variables. Thus, as expected, the number of nodes in the input layer differs in accordance with the number of parameters involved in each model. Consequently, the capability (correct classification) of each model to correctly classify the soil liquefaction potential, using available inputs contained within the data sets, varied. For brevity, only the three most promising stand-alone models (designated herein as model A, B and C) along with the combined model (i.e., ensemble of models B and C) are presented in this paper.

Model A

This model utilizes four input parameters to classify the soil liquefaction potential. Required input parameters for this model are: earthquake magnitude, M, effective stress on the susceptible layer, σ'_v, uncorrected SPT value, N and the epicenter distance, r. The output layer is made up of one node describing the liquefaction potential status. In this case, 1 is used to represent a liquefied case while 0 is used to represent a non-liquefied case. Upon training and testing on 44 and 17 data sets, respectively, this network was found to yield the lowest percentage of misclassification in prediction for both training (recall) and testing sets (on-line validation) at 3 hidden nodes. Consequently, further validation of this network was performed on the remaining 44 data sets which had never been used in either training or testing phases. The number of misclassification cases were 4, 0 and 15 for training, testing and validation data sets considered,

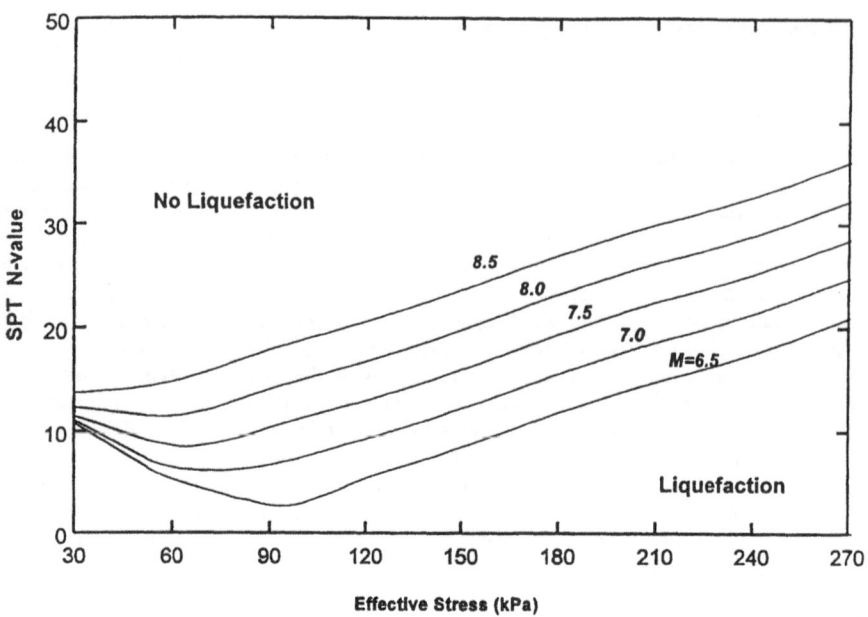

Figure 3 : Liquefaction potential assessment chart based on BPNN-model B

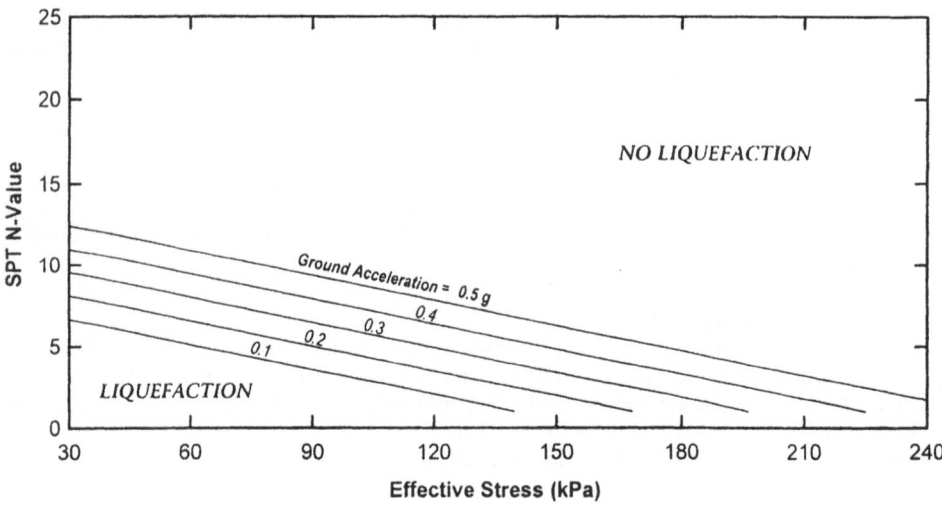

Figure 4 : Liquefaction potential assessment chart based on BPNN-model C

respectively. These numbers represent misclassification percentages of about 9%, 0% and 34% for training, testing and validation cases, respectively. This results in an overall misclassification percentage of about 18%. Therefore, this network may yield about 82% correct classification using only the validation data sets truly represents the accuracy of the developed model, then the percentage of correct classification would drop to about 66%.

In order to make use of this model by other geotechnical engineers, the developed BPNN model was used to assess the liquefaction potential for given input parameters. This procedure was repeated for hundreds of times by varying M, σ'_v, uncorrected N and r values within the applicable ranges observed in the original database. All input parameter values considered and corresponding liquefaction potential values were complied to produce the 0.5 liquefaction potential contour lines that represent the boundary between liquefaction and non-liquefaction zones. These contour lines were then superimposed to produce a set of liquefaction assessment charts for various values of earthquake magnitudes. A sample of these charts for M=7.0 is depicted in Fig. 1. Contour lines shown in this figure represents the boundaries of $(r_{max})_{critical}$ zones. For example, for given values of M, σ'_v, uncorrected N and r, if the case plots below the specified $(r_{max})_{critical}$ line, then this case indicates that liquefaction potential is favorable. Alteratively, if the case plots above its corresponding $(r_{max})_{critical}$ line, then this case indicates that liquefaction potential is not favorable.

Model B

This model utilizes three input parameters to classify the soil liquefaction potential. Required input parameters for this model are: earthquake magnitude, M, effective stress on susceptible layer, σ'_v, and uncorrected SPT value. This model assumes that the location of the site in question lies within the influence zone of the earthquake under consideration. For this reason, the epicenter distance variable was eliminated from the input layer. Similar to model A, the output layer is made up of one node describing the liquefaction potential status where 1 is used to represent the liquefied case while 0 is used to represent the non-liquefied case. Upon training and testing on 44 and 17 data sets, respectively, this network was found to yield the lowest percentage of misclassification in prediction for both training and testing sets at 2 hidden nodes. Consequently, further validation of this network was performed on the remaining data sets. The number of misclassification cases were 5, 2 and 8 for training, testing and validation sets considered, respectively. This results in an overall misclassification percentage of 14%. In other words, this network may yield about 86% correct classification accuracy if all sets are considered. On the other hand, if we, conservatively, assume that the percentage of correct classification using only the validation sets truly represents the accuracy of the developed model, then the percentage of correct classification will slightly drop to about 82%.

Due to the significant advantages of this model (i.e., having both least input parameters and high degree of prediction accuracy) it was utilized to produce the liquefaction potential assessment map shown in Fig.2. This map was produced from the actual output of BPNN model B by varying the values of the SPT and the effective stresses, while keeping $M = 7$. All output results were used to draw the 0.5 contour line which is considered herein to represent the 50% probability for either liquefaction or non-liquefaction potential (i.e. the divide between liquefaction and non-liquefaction zones) as depicted in Fig.2. To illustrate the accuracy of this chart. Based on this overlaying, it was found that only five misclassifications are reported out of the 35 cases considered which represent about 85% classification accuracy.

By using the same approach, potential liquefaction values were obtained using this model for $M = 6.5$, 7.0, 7.5, 8.0 and 8.5 on the same graph and then plotting only the 0.5 contour line for each case yields the family of curves shown in Fig. 3. The curves shown in this figure represent the boundary between liquefaction and non-liquefaction potential zones for different earthquake magnitudes. For each curve, the area to the bottom-right represents the liquefaction zone, while the area to the top-left represents the non-liquefaction zone. As an illustrative example, for $M = 6.5$, $N = $ (SPT) = 20 and $\sigma'_v = 200$kPa, according to Fig. 3, liquefaction potential is not favorable. If the value of M is to increase to 7.5 or above, then the liquefaction potential becomes more favorable.

Model C

Similar to model B, this model utilizes only three input parameters to classify the soil liquefaction potential. The only difference between this model and model B is the utilization of the maximum horizontal ground acceleration, a_{max} variable as an input parameter instead of the end of the earthquake magnitude variable. In this case, the other two input parameters (effective stress and SPT value) were kept unchanged. As a result of this change, the required input parameters for this model are: maximum horizontal ground acceleration, a_{max}, effective stress on the susceptible layer, σ'_v, and uncorrected SPT N-value. Additionally, the utilization of a_{max}

instead of M in the input layer, eliminates the need for assuming that the location of the site in question lies within the influence zone of the earthquake under consideration. Based on the input parameters involved in each of the previous models, model C can be referred to as a local model while models A and B are referred to as source models. Similar steps of training and testing utilized in model B were also used to develop model C. Upon training and testing on similar number of data sets, this network was found to yield the lowest percentage of misclassification at 1 hidden node. Unfortunately, further validation of this network was not possible since the values of a_{max} for all 44 validation sets used in models A and B were missing. Therefore, only number of misclassification cases for training and testing sets can be evaluated. Based on this information, the model yields classification percentages accuracy of about 73% and 76% for training and testing cases, respectively. This represents an overall classification percentage accuracy of about 74%.

Due to the small number of parameters involved in this model and the modest degree of prediction accuracy, it was utilized to produce the liquefaction potential assessment map shown in Fig. 4. Contour lines shown in this figure represents the boundaries of $(a_{max})_{critical}$ zones. For example, for given values of σ'_v and uncorrected N, if the case plots below the specified $(a_{max})_{critical}$ line, then this case indicates that liquefaction potential is not favorable. Alternatively, if the case plots above its corresponding $(a_{max})_{critical}$ line, then this case indicates that liquefaction potential is not favorable. Further examination of this chart reveal the possibility of deducing this chart to simple and useful equation that can aid geotechnical engineers in liquefaction assessment tasks. Nonetheless, further validation of this chart/deduced equation on new data sets is needed.

Ensemble of models B and C

In this model, the trained networks of model B and c are used simultaneously to aid in arriving at the final decision about liquefaction or non-liquefaction potential. The steps involved in using this ensemble model are listed below:

Step 1: Use model B (i.e., BPNN-based model or chart depicted in Fig. 3) to obtain the liquefaction potential value.

Step 2: Use model C (i.e., BPNN-based model or chart depicted in Fig. 4) to obtain the liquefaction potential value.

Step 3: Apply the applicable case (from those explained below) and make the final decision in regard to liquefaction potential assessment possibility.

According to this model, only one of the following cases should be applicable for a specific situation.

Case 1: If both models forecast liquefaction possibilities (i.e., 1;1), then this full-agreement situation is categorized as an overall liquefaction case.

Case 2: If both models project non-liquefaction possibilities (i.e., 0, 0), then this full-agreement condition is categorized as an overall non-liquefaction case.

Case 3: If model B predicts a non-liquefaction situation while model C predicts a liquefaction situation (i.e., 0, 1), then no decision can be made in this situation. This case indicates that further detailed examination and/or analysis are warranted.

Case 4: If model B projects a liquefaction state while model C projects a non-liquefaction state (i.e., 1, 0), then no decision can be made under this scenario. This case also indicates that further detailed examination and/or analysis are warranted.

According to this approach, if both predictions of models B and C conform to case 1 or 2 and agrees with actual observation, then this is considered a success classification case. On the other hand, if both predictions of the models conform to case 1 or 2 and disagrees with actual observation, then this is considered a failure classification accuracy compared with the one observed using any of the previously mentioned stand-alone models (i.e., model A, B or C). It is the opinion of the authors that when models B and c disagree on the outcome (case 3 and 4), then this predicament should be considered as a signal for the need of further detailed examination of the situation at hand by more sophisticated techniques.

Table 1: Number of misclassification cases yielded by all models considered in this study.

| Model | Seed & Idriss (1971) | Berrill & Davis (1985) | Fuzzy Logic* 2-D | | Fuzzy Logic* 3-D | | BPNN-based Models | | |
			Seed	Berrill	Seed	Berrill	Model A	Model B	Model C
61 records	21	12	17	11	15	10	4	7	16
105 records	NA	28	NA	24	NA	NR	19	15	NA

NR: Not Reported   NA: Not Applicable   After El Zahaby (1995)

Comparison with Other Methods:

Three methods representing different approaches [i.e., cyclic stress analysis (Seed and Idriss, 1971), energy dissipation principle (Berrill and Davis, 1985) and fuzzy logic approach (El Zahaby, 1995) to determine liquefaction potentials are used herein for direct comparative purposes with BPNN-based models A, B and C. It is to be noted that the prediction accuracy of those three methods was reported by El Zahaby (1995). The complete comparison of the performance of all models considered herein is given in Table 1. It is obvious from table from Table 1 that BPNN-based models A and B yield the best prediction accuracy compared to other methods - the reason that model C was less accurate than models A or B is mainly due to the utilization of the maximum peak horizontal acceleration, a_{max} variable as a key input parameter to the model. Generally speaking, evaluation of this input parameter involves some degree of uncertainty in comparison with the earthquake magnitude variable used in models A and B.

Using the first group (61 records), the number of misclassifications using Seed & Idriss (1971) model was 21 while, in the case of Berrill & Davis model (1985), the number of misclassified cases was reduced to 12. Applying the 2D fuzzy logic technique, developed by El Zahaby (1995), on both models enhanced the number of misclassifications in Seed & Idriss model (1971) to 17, while for Berrill & Davis (1985) model the number was slightly reduced from 12 to 11. Incorporating the 3-D fuzzy analysis reduced the number of misclassifications to 15 and 10, respectively. On the other hand, applying the BPNN technique, the number of the misclassifications was reduced to 7 and 4 when models A and B were used, respectively. This means that the BPNN approach was able to reduce the number of misclassification by more than 21% in comparison with the enhanced results obtained from the Berrill and Davis (1985) model with inclusion of 2_fuzzy logic. It should be noted that Seed and Idriss (1971) model and BPNN-based model C could not be used on the third group data sets (44 validation records) due to the missing of some key input parameters in these records. Also, El Zahaby (1995) did not perform a 3-D analysis on those additional 44 records.

CONCLUDING REMARKS

Based on the results and findings obtained so far by the KSU research team in regard to the use of BPNN in the soil liquefaction potential assessment area, the following remarks can be highlighted.

1. BPNN approach was successfully used to assess the soil liquefaction potential by using data obtained from previous case histories. According to the results obtained herein, BPNN-based models were able to produce the most accurate assessment for liquefaction potential compared to assessments obtained using other techniques investigated in this study.

2. The ensemble model approach should be used when possible since it combines the knowledge of more than one model and at the same time offers various unique features in regard to flexibility in interpreting the final classification outcome.

3. BPNN-Model B was found to be the most accurate stand-alone model which requires only the use of a small number of easily obtainable earthquake and site-related parameters without the need for any complicated analysis or intricate laboratory tests.

4. BPNN-based models were able to capture or realize implicitly the inter-relation between the effective stress and the uncorrected SPT values.

REFERENCES

Agrawal,G., Chameau, J.L., & Bourdeu, P.L. 1997. Assessing the Liquefaction Susceptibility at a Site Based on Information from Penetration Testing. Artificial Neural Networks for Civil Engineers: Fundamentals and Applications, ASCE, pp 185-214.

Ali, H.E. & Najjar, Y.M. 1998. Neuronet-Based Approach for Assessing the Liquefaction Potentials of Soils. (in press) Transportation Research Board.

Berrill, J.B. & Davis, R.O. 1985. Energy Dissipation and Seismic Liquefaction of Sands: Revised Model. Soils and Foundations, Vol. 25, No.2, pp. 106-118.

Christian, J.T. & Swiger, W.F. 1975. Statistics of Liquefaction and SPT Results, Journal of Geotechnical Engineering, Vol. 101, No. GT11, pp. 1135-1150.

Ellis, G. W., Yao, C., Zhao, R., & Penumadu, D. 1995. Stress-Strain Modeling of Sand using Artificial Neural Network. Journal of Geotechnical Engineering, Vol. 121, No. 5, pp. 429-435.

Elton, D.J., Juang, C.H., & Sukumaron, B. 1995. Liquefaction Susceptibility Evaluation Using Fuzzy Sets. Soils and Foundations, Vol. 35, No. 2, pp. 49-60.

El Zahaby, K.M. 1995. Liquefaction Risk Analysis Including Fuzzy Variables. Ph.D. dissertation, North Carolina State University.

El Zahaby, K.M. & Rahman, M.S. 1996. Non-Statistical Uncertainities in Liquefaction Risk Assessment. Uncertainity in the Geologic Environment: From theory to practice, C.D. Shackelford et al (Ed.), Vol. 2, pp. 1068-1082.

Goh, A.T.C. 1994. Seismic Liquefaction Potential Assessed by Neural Networks. Journal of Geotechnical Engineering, Vol. 120, No. 9, pp. 1467-1480.

Hassoun, M. 1995. Fundamentals of Artificial Neural Networks. MIT Press, Cambridge, Mass.

Kramer, Steven K. (1996). Geotechnical Earthquake Engineering. Prentice-Hall, Inc.

Najjar, Y.M. and Ali, H. E. (1998). SPT-Based Liquefaction Potential Assessment: An artificial Neural Network Approach. Proceedings, 12th Engineering Mechanics Conference, La Jolla, CA. May 17-20, Murakami & Luco (Editors), pp. 960-964.

Najjar, Y.M., and Basheer, I.A. 1996. Utilizing Computational Neural Networks for Evaluating the Permeability of Compacted Clay Liners. Geotechnical and Geological Engineering, 14, pp. 193-212.

Najjar, Y.M., Basheer, I.A., & Naouss, W.A. 1996. On the Identification of Compaction Characteristics by Neuronets. Computer and Geotechnics, Vol. 18, No. 3, pp. 167-187.

Seed, H.B. and Idriss, I.M. 1971. Simplified Procedure for Evaluating Soil Liquefaction Potential. Journal of Soil Mechanics and Foundations Division, ASCE, Vol. 97, No. SM9, pp. 1249-1273.

Seed, H.B. & Idriss, I.M., & Arango, I. 1983. Evaluation of Liquefaction Potential Using Field Performance Data. Journal of Geotechnical Engineering, Vol. 109, No. 3, pp. 458-482.

Seed, H.B., Tokimatsu, K., Harder, L.F., & Chung, R.M. 1985. Influence of SPT Procedures in Liquefaction Resistance Evaluation. Journal of Geotechnical Engineering, Vol. 111, No. 12, pp. 1425-1445.

compared with the sigmoid activation function. While in the connection between the hidden layer and the output layer, a linear transform function is adopted for the sake of simplicity.

(3) Training process: The training process is initialized by inputting a set of "comprehensive" data to the network. The so called "comprehensive" data imply that the data are most representative and contain all the necessary information for the given problem. A network trained with the comprehensive data can have strong predictability. The flexibility of the network itself ensures that the network can be trained ceaselessly until the comprehensive data are included in the training data base. As indicated in Table l, the test data numbered as C1, C3, C4 and C8 are used as training data since they contain possible range of the data. Each data set contains about 1100 data. It is necessary, from our experience, to normalize every set of data

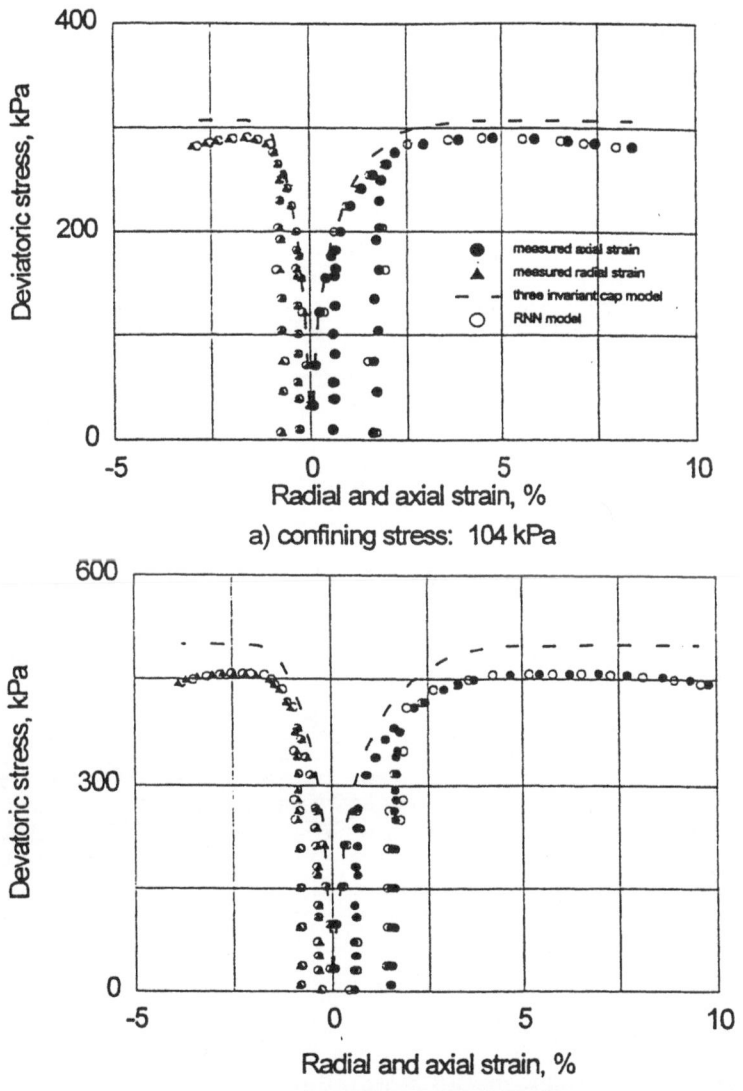

a) confining stress: 104 kPa

b) confining stress: 172 kPa

Figure 3   Simulation of stress-strain relationships of experimental data with the RNN model together with the three invariant-based cap model

Artificial Intelligence and Mathematical Methods in Pavement and Geomechanical Systems, Attoh-Okine (ed.)
© *1998 Balkema, Rotterdam, ISBN 90 5809 028 0*

A system of neural networks for deriving soil parameters from in-situ tests

C. Hsein Juang
Department of Civil Engineering, Clemson University, S.C., USA

S. H. Ni & P. C. Lu
Department of Civil Engineering, National Cheng Kung University, Tainan, Taiwan

ABSTRACT: Deriving soil parameters from in-situ tests is usually the first step in the design of a geotechnical system. This paper is intended to show that neural network technology can be an effective tool for deriving soil parameters from in-situ tests. As an example to support this viewpoint, a system of artificial neural networks (ANN) is developed for predicting sand parameters such as relative density (D_r), coefficient of lateral earth pressure at rest (K_0), and overconsolidation ratio (OCR) from cone penetration test (CPT), a widely-used in-situ test. This paper focuses on the results pertain to the estimation of Dr using ANN.

1 INTRODUCTION

Many empirical equations exist for determination of geotechnical parameters such as relative density (D_r), coefficient of lateral earth pressure at rest (K_0), and overconsolidation ratio (OCR) of sands based on cone penetration test (CPT). These empirical equations are mostly developed based on CPT calibration chamber data using statistical regression methods. Indeed, quality calibration chamber data around the world are now more abundant than before. However, the accuracy of the existing equations for predicting geotechnical parameters such as K_0 and OCR is often less than desirable. It would be of interest to the geotechnical engineer to develop new methods that are more accurate than the existing methods in light of the availability of more data and the recent advance in the area of data analysis techniques.

Least-square regression is a powerful technique that the geotechnical engineer has relied upon for establishing empirical equations for many decades. One problem with the regression method, however, is that it requires a model (i.e., the form of the regression equation) to begin with. For example, if it is desired to establish a model for determining K_0 from CPT measurements, one would probably try different regression models taking K_0 as the dependent variable and CPT parameters as the independent variables. Since a prior knowledge of the model is required, it is often difficult to choose the "right" model for conducting the regression analysis, *if the input/output relationship is highly non-linear and complex*. No systematic guidance is available to search for the right model. To overcome this problem, artificial neural network (ANN) approach is taken in the present study. Artificial neural networks, referred to herein simply as neural networks or networks, can map input to output without prior knowledge of the underlying physical model. ANN has the ability to *learn* from examples. Because of this characteristic, appropriate "architecture" or model of the neural network can be obtained through a systematic search and evaluation. Use of ANN to approximate cause-effect relationships in geotechnical engineering has been reported in recent years (for example, Agrawal, *et al.*, 1995; Juang & Elton, 1997; Juang & Chen, 1998; Juang, *et al.*, 1998; Ni, *et al.*, 1996).

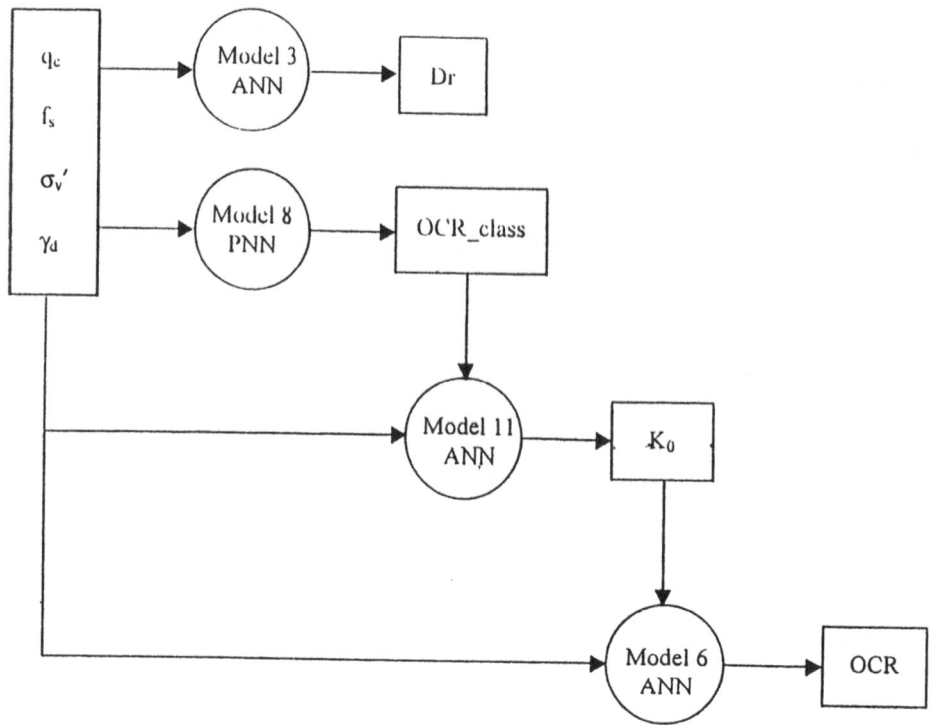

Figure 1 Proposed system for determining D_r, K_0, and OCR of sands

2   DEVELOPMENT OF THE PROPOSED SYSTEM

As an example to show the neural network's capability in performing the task of function approximation and classification, a system is developed for predicting design parameters of sands such as D_r, K_0, and OCR from CPT data. The system consists of several neural networks as illustrated in Figure 1. This paper focuses on the issue of D_r prediction.

2.1   *Source Data Used in the Development of Neural Networks*

The data used in the present study come from two sources. One is the collection of the results of calibration chamber tests in Ticino and Hokksund sands (Lunne, *et al.*, 1997). The other is the collection compiled by Sagaldo (1993).

Each "data point" of the data set used in the present study consists of the following information: dry unit weight after consolidation of sand sample (γ_d), relative density after consolidation of sample (D_r), applied vertical consolidation stress (σ'_v), lateral stress divided by horizontal stress (K_0), overconsolidation ratio (OCR), measured cone resistance (q_c), measured sleeve friction (f_s), cone penetrometer diameter (d_c), and boundary condition (BC).   One hundred and sixty eight (168) data points are available from Lunne's collection and one hundred and seventy one (171) data points are available from Sagaldo's collection. Each data point is called an *instance* in the present study, and a total of 339 instances is used. Among these data instances, 204 instances are used for training neural networks, 101 instances are used for testing the trained networks, and 34 instances are used for additional validation.

2.2   *Neural Network for Predicting D_r*

Existing Empirical Methods

Several empirical methods are available in the literature for estimating D_r. Robertson & Campanella (1983) presented a chart that relates D_r to q_c and σ'_v for normally consolidated sands. The influence of sand compressibility is also indicated in their chart, although no quantitative measure of the compressibility is specified. Juang, *et al.* (1996) presented a fuzzy-set-based method that takes into account of sand compressibility in the formulation. In addition to the two major input parameters, q_c and σ_v', the sleeve friction (f_s) was introduced into their formulation to account for the effect of sand compressibility. Baldi, *et al.* (1986) published an empirical equation that relates D_r to q_c, σ_v', and K_0. Kulhawy & Mayne (1991) proposed yet another empirical equation that relates D_r to q_c, σ_v', and OCR.

Selection of Input Variables

The rationale for selection of input variables for neural networks development in the present study is described in the following. D_r of a sand, with no prior knowledge of whether it is normally consolidated or overconsolidated, may be considered as a function of input parameters q_c, f_s, σ_v', OCR, and K_0. However, OCR and K_0 are unknown in a CPT test. In fact, they are often the parameters that are to be "predicted" based on in-situ tests, and thus should be excluded from the list. On the other hand, the vertical total stress (σ_v), while not used in the aforementioned empirical equations, is incorporated in a comprehensive CPT-based method for evaluating liquefaction resistance (Robertson & Wride, 1997). In the present study, the dry unit weight of sand (γ_d), which is available in the data set, is selected in place of σ_v. Use of the variables σ_v' and γ_d is thought to yield comparable effects as that would be achieved using the variables σ_v' and σ_v. The ultimate decision as to which parameters to include in the neural network models, however, rests on the performance of the trained neural networks. Thus, the following ANN models are "tried" in the present study:

Model 1: $D_r = f(q_c, \sigma_v')$
Model 2: $D_r = f(q_c, f_s, \sigma_v')$
Model 3: $D_r = f(q_c, f_s, \sigma_v', \gamma_d)$

Normalization of Data for Network Training and Testing

The source data were described earlier. Normalization of the factors (or variables) in an instance (input/output pair) is generally required before feeding data to the ANN models. In the present

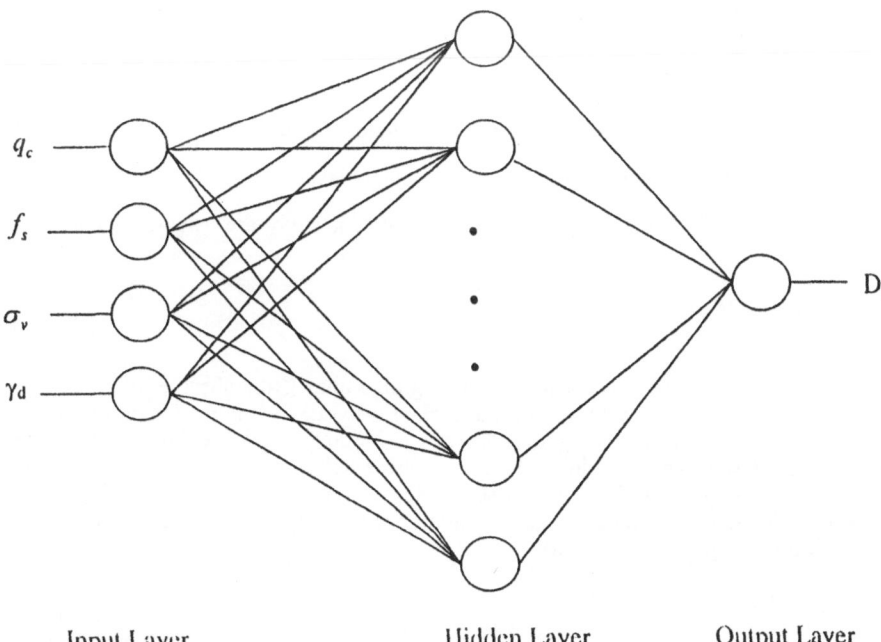

Input Layer   Hidden Layer   Output Layer

Figure 2   Elements of a three-layer artificial neural network

study, all input/output variables are normalized with respect to the minimum and maximum of the data into the range of 0.1 to 0.9. Symbolically, the normalization procedure may be expressed as:

$$x_{norm} = \frac{x + a}{b}$$

where $\quad a = (x_{max} - 9x_{min})/8$

$\qquad\quad b = (x_{max} - x_{min})/0.8$

and where x is the actual parameter value, x_{min} and x_{max} are the minimum and maximum value, respectively.

Network Topology and Training Algorithm

A three-layer, feed-forward network topology, as shown in Figure 2, is adopted in this study. The objective of the network training is to map the input to the output by determining the connection weights and biases through an error reduction technique. For the three-layer network shown in Figure 2, the network prediction z (or D_r in Models 1, 2, and 3) is calculated as follows:

$$z = f_2\left\{ B_0 + \sum_{k=1}^{n}\left[W_k \cdot f_1\left(B_{Hk} + \sum_{i=1}^{m} W_{ik}P_i \right)\right]\right\} \tag{2}$$

where $\quad B_0 \quad$ = bias at the output layer (just one neuron in this layer),

$\qquad\quad W_k \quad$ = weight of connection between neuron k of the hidden layer and the single output layer neuron,

$\qquad\quad B_{Hk} \quad$ = bias at neuron k of the hidden layer ($k = 1$, n),

$\qquad\quad W_{ik} \quad$ = weight of connection between input variable i ($i = 1$, m) and neuron k of the hidden layer,

$\qquad\quad Pi \qquad$ = input parameter i,

$\qquad\quad f_1(\lambda) \quad$ = transfer function of each neuron in the hidden layer, and

$\qquad\quad f_2(\lambda) \quad$ = transfer function of the neuron in the output layer.

Both transfer functions $f_1(\lambda)$ and $f_2(\lambda)$ adopted in the present study are "sigmoid" functions defined below:

$$f_N(\lambda) = \frac{1}{1 + e^{-\lambda}} \quad \text{for N} = 1,2 \tag{3}$$

In the present study, the back-propagation learning algorithm (Rumelhart, et al, 1986), an error reduction technique, is adopted to train the desired network for the prediction of Dr. As shown in Figure 2, the connection weights and biases to be determined are those between the neurons in the input layer and the hidden layer, and those between the hidden layer and the output layer. The appropriate weights and biases are to be "learned" from available instances. The aim of the training (or learning) process is to minimize the global network error, defined as (for networks with a single output neuron):

$$E = \frac{1}{2}\sum_{j=1}^{p}\left(z_j - o_j \right)^2 \tag{4}$$

where p is the number of instances used in the training, z_j is the network prediction (i.e., actual

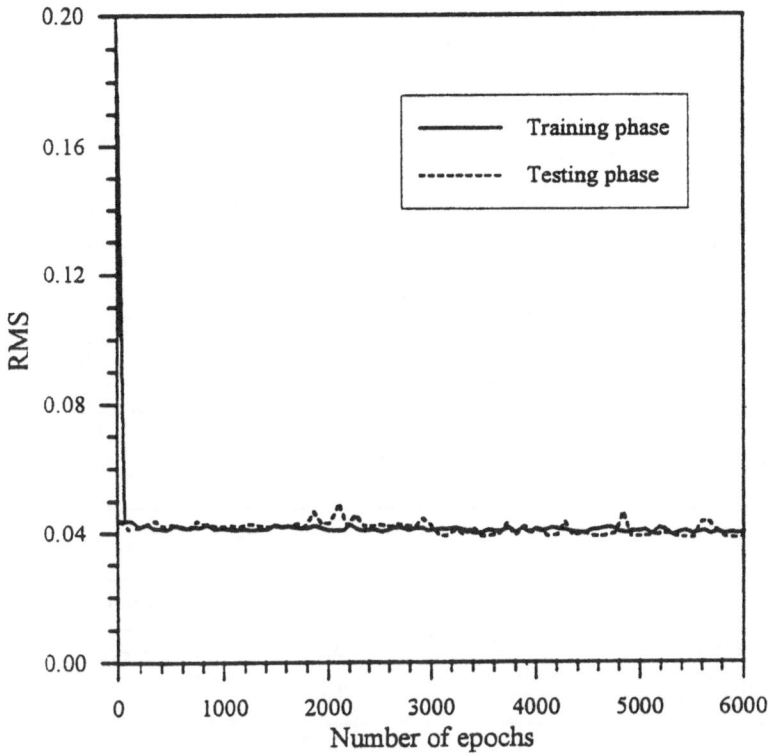

Figure 3 Reduction of network error in the development of Model 3-based ANN

network output), and oj is the target output. In the back-propagation training process, the network error is back propagated into each neuron in the hidden layer, and then continued into the neurons in the input layer. The distributed error at each neuron is the basis for modification of connection weights and biases. By continuous modifications of the connection weights and biases, the global network error is gradually reduced. In practice, an error goal is set before the network training, and if the network error during the training becomes less than the error goal, the training is stopped.

For each of Models 1, 2, 3 described above, the number of neurons in the hidden layer is determined through a trial-and-error process, as is normally done in back-propagation neural network training. Optimal number of neurons is often difficult to define. However, the smallest number of neurons that are required to yield "satisfactory" results (judged by the network performance) is usually preferred. Because prior knowledge of an underlying physical model is not required, the training of the networks is indeed a systematic and mechanical process.

Results – Prediction of Dr

Figure 3 show the reduction of network error in the training and testing processes using Model 3. Here, the root-mean-square (RMS) of errors of all instances are shown for both training and testing. For this particular problem, the converged results were achieved at about 100 epochs (note: one complete presentation of the entire set of training patterns during the training process is an epoch). Additional training of the network until 6000 epochs produced essentially the same results. The actual run time for 6000 epochs was about an hour using an Intel Pentium 166 MHz processor-based PC. These results were deemed acceptable, as the RMS in both training and testing is relatively small.

Table 1 shows the performance of the trained ANNs in predicting Dr for all three models. For each model, the results shown in Table 1 represent the best performing network that can be

69

Table 1 Performance of ANN Models 1, 2, and 3 in predicting D_r

Model	Inputs	No. of nodes in hidden layer	Standard error of predicted D_r		R^2 in predicted vs. measured D_r		Success rate (%)*	
			training	*testing*	*training*	*testing*	*training*	*testing*
1	q_c, σ'_v	10	9.3	8.3	0.79	0.83	79.4	84.2
2	q_c, f_s, σ'_v	8	8.7	8.1	0.80	0.83	80.4	85.2
3	q_c, f_s, σ'_v, γ_d	6	5.1	5.1	0.95	0.95	90.7	92.1

* Note that the success is defined here as being able to predict D_r within an error of ± 20%.

Table 2 Comparison of existing empirical methods with proposed system

Method	Success Rate (%)		
	D_r	K_0	OCR
Eq 2. - Baldi, et al (1986)	53.7	5.3	N/A
Eq 5. - Kulhawy & Mayne (1991)	83.5	N/A	19.8
Eq 18. - Mayne (1991)	N/A	48.7	34.5
Proposed system	90.9	91.2	93.2

achieved in the present study. Among the three models, Model 3 yields the best results, in terms of the standard error, the coefficient of determination (R2), and the success rate. It is noted that all four input parameters of Model 3 are generally available in CPT-based site characterization.

Note that the data used in the training and testing in the above discussion consist of both normally consolidated and overconsolidated sands. Thus, the knowledge of OCR or K0 is not required when using Model 3-based ANN for determining Dr. Also note that the results of Model 3-based ANN can be easily "ported" to a spreadsheet program and thus it is rather easy to use in practice.

3 COMPARISON WITH EXISTING EMPIRICAL METHODS

Figure 4 shows a comparison of D_r predictions for three methods, the Baldi method (Baldi, *et al.*, 1986), the Kulhawy method (Kulhawy & Mayne, 1991), and the proposed system. Note that with the Baldi method, the value of K_0 is assumed known. In other words, D_r is calculated given the benefit of knowing the actual K_0 values. With the Kulhawy method, the effects of compressibility and aging are not considered. However, as with the Baldi method, this method is given the benefit of knowing the actual OCR in the calculation of D_r. On the other hand, the proposed system does not use K_0 or OCR, as both are generally not available in CPT measurements. As shown in Figure 4, the proposed system produces the most accurate estimate of D_r, even it does not have a prior knowledge of OCR or K_0.

As previously shown in Figure 1, the proposed system consists of a neural network for predicting D_r, a neural network for classifying sands into different OCR classes, a neural network for predicting K_0, and a neural network for predicting OCR. Because of space limit,

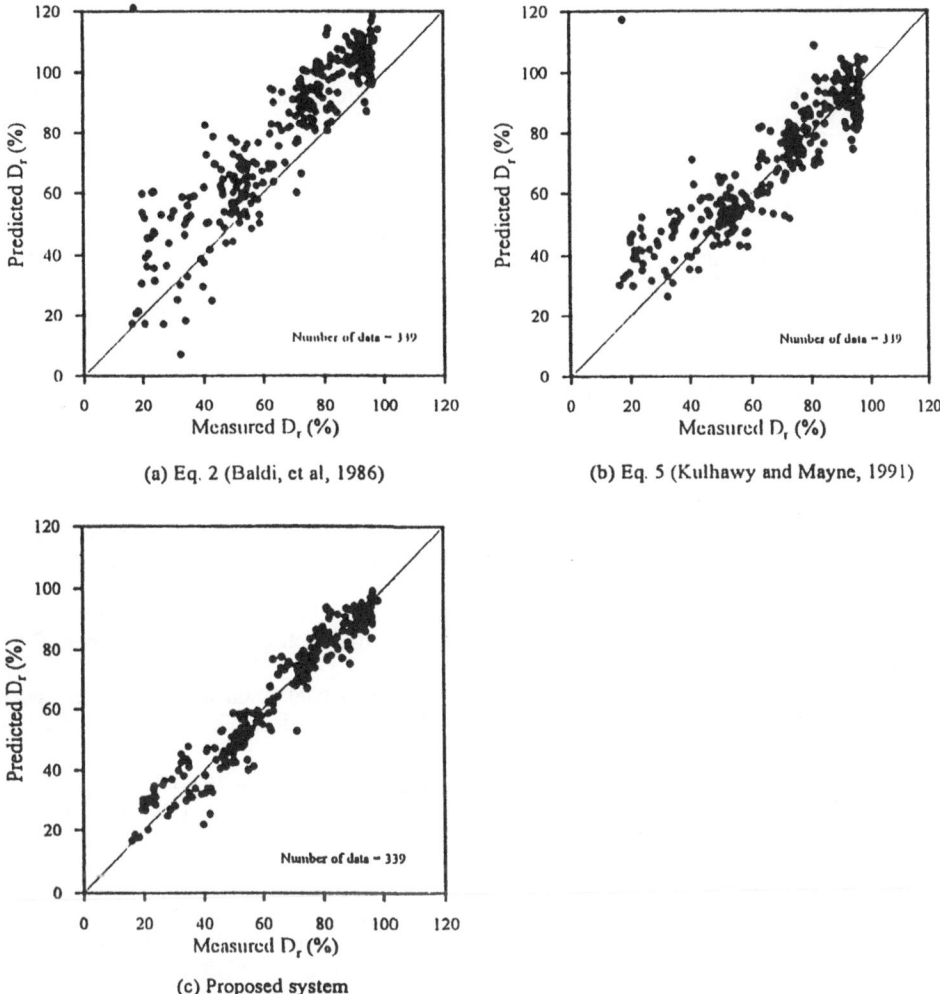

Figure 4 Comparison of existing methods with proposed system for prediction of D_r

detail on the last three neural networks shown in Figure 1 is not presented here. However, Table 1 shows a summary of comparisons of the accuracy of the ANN predictions versus those obtained from existing, regression-based methods. Here, the accuracy of a particular parameter prediction is measured by success rate, where a prediction is considered a success if the percent error between the prediction and the actual parameter value is less than a certain percentage, say, 15%. The superior predicting capability and accuracy of the proposed system are evidenced.

4   CONCLUSIONS

The proposed system as documented in this paper shows the potential of neural networks as a tool for establishing empirical methods for predicting geotechnical parameters.
The proposed system is shown far superior to existing regression-based empirical methods for predicting D_r, K_0, and OCR of sands based on CPT data.

ACKNOWLEDGMENTS

The study on which this paper is based is supported by the National Science Foundation through Grant No. CMS-9610103. The cognizant NSF program official for this grant is Dr. Priscilla P. Nelson. This financial support is greatly appreciated.

REFERENCES

Agrawal, G., Chameau, J.L., & Bourdeau, P.L., 1995. "Assessing the liquefaction susceptibility at a site based on information from penetration testing." Chapter 10, in: Artificial Neural Networks for Civil Engineers – Fundamentals and Applications, ASCE Monograph, New York.

Baldi, G., Bellotti, R., Ghionna, V., Jamiolkowski, M., & Pasqualini, E., 1986. "Interpretation of CPT's and CPTU's. 2nd part: drained penetration on Sands," Proc., 4th Int. Geotech. Seminar: Field Instrumentation and In-Situ Measurements, Nanyang Technol. Univ., Singapore, pp. 143-162.

Demuth & Beale., 1997. Neural Network Toolbox, User's Guide, Version 3, The Math Works, Inc., Natick, MA 01760.

Juang, C.H., Huang, X.H., Holtz, R.D. & Chen, J.W., 1996. "Determining relative density of sands from CPT using fuzzy sets," Journal of Geotechnical Engineering, Vol. 122, No. 1, pp. 1-6.

Juang, C.H. & Elton, D.J., 1997. "Predicting collapse potential of soils with neural networks," Transportation Research Record, No. 1582, National Research Council, Washington, D.C., pp. 22-28.

Juang, C.H. & Chen, C.J., 1998. "CPT-based liquefaction evaluation using artificial neural networks," accepted for publication, Journal of Computer-Aided Civil and Infrastructure Engineering, April, 1998, Blackwell Publishers, Cambridge, MA.

Juang, C.H., Ni, S.H., & Lu, P.C., 1998. "Training artificial neural networks with the aid of fuzzy sets," accepted for publication, Journal of Computer-Aided Civil and Infrastructure Engineering, April, 1998, Blackwell Publishers, Cambridge, MA.

Kulhawy, F.H. & Mayne, P.W., 1991. "Relative density, SPT, and CPT Interpretations," Proc., 1st Int. Symposium on Calibration Chamber Testing, Potsdam, New York, Editor, A.B. Huang, Elsevier, Amsterdam, pp. 197-211.

Lunne, T., Robertson, P.K., & Powell, J.J.M., 1997. Cone Penetration Testing in Geotechnical Practice, Blackie Academic & Professional, London, ISBN. 0-751-40393-8.

Mayne, P.W. (1991), "Tentative method for estimating $\sigma ho'$ from qc data in sands," Proc., 1st Int. Symposium on Calibration Chamber Testing, Potsdam, New York, Editor, A.B. Huang, Elsevier, Amsterdam, pp. 249-256.

Mayne, P.W. & Kulhawy, F.H., 1991. "Calibration chamber database and boundary effects correction for CPT data," Proc., 1st Int. Symposium on Calibration Chamber Testing, Potsdam, New York, Editor, A.B. Huang, Elsevier, Amsterdam, pp. 257-264.

Ni, S. H., Lu, P. C., & Juang, C. H. (1996), "A fuzzy neural network approach to evaluation of slope failure potential," Microcomputers in Civil Engineering, Blackwell Publishers, Cambridge, Mass., Vol. 11, pp. 59-66.

Robertson, P.K. & Campanella, R.G., 1985. "Liquefaction potential of sands using the cone penetration test," Journal of the Geotechnical Engineering Division, ASCE, Vol. 111, GT 3, 1985, pp. 384-403.

Robertson, P.K. & Wride, C.E., 1997. "Cyclic liquefaction and its evaluation based on SPT and CPT," Geotechnical Group, University of Alberta, Edmonton, Alberta, Canada, 1997, 52pp.

Rumelhart, D.E., Hinton, G.E., & Williams, R.J. (1986), "Learning internal representations by error propagation," In D.E. Rumelhart and J. L. McClelland, eds., Parallel Distributed Processing, Vol. 1, Chap. 8, MIT Press, Cambridge, MA, pp. 318-362.

Sagaldo, R., 1993. "Analysis of Penetration Resistances in Sands," Ph.D. thesis, University of California at Berkeley, California.

Artificial Intelligence and Mathematical Methods in Pavement and Geomechanical Systems, Attoh-Okine (ed.)
© *1998 Balkema, Rotterdam, ISBN 90 5809 028 0*

Artificial immune systems and networks for pavement deterioration process

N.O. Attoh-Okine
Department of Civil Engineering, Florida International University, Miami, Fla., USA

ABSTRACT: Artificial Immune Systems (AIS) are adaptive systems based upon the models of natural systems in which learning takes place by evolutionary mechanism in a similar way to biological evolutions. The immune system, whose functions is to identify and eliminate foreign material, is capable of learning, memorizing and recognizing patterns. This paper presents the basic theory of AIS and identifies the potential application of AIS and immune networks to the pavement deterioration process.

1 INTRODUCTION

Pavement deterioration models are used to estimate the life-cycle impacts of investments in transportation facilities in terms of costs and benefits and the condition of the facility. Types of deterioration models include: (1) Regression models- Condition (as a damage aggregate measure) is modeled as a function of exogenous or casual variables, such as traffic, environment, facility construction and characteristics, and maintenance and rehabilitation actions. Examples of this approach include aggregate or disaggregate condition measures as a function of usage, strength and environment (Paterson & Attoh-Okine 1992). Regression models are easy to interpret but they do not explicitly deal with errors in the data or functional form, (2) Markov models assume that the probability of a facility to change from one state to another is only dependent on its current state (Feighan *et al.* 1987), (3) Survivor curves represent the percentage of facilities that remain in service as a function of time, (4) Latent variable models assume that the true condition of the facility is unobserved and that measures are indicators of damage. The models consist of casual equations representing the relationship between the latent variable causing deterioration and measurement equations for each indicator (Ben-Akiva *et al.* 1993). The disadvantage in this approach is that it is too subjective, (5) Two-phase models look at individual damage indicators consisting of initiation phase and progression (Paterson 1987). The disadvantage is that not all damage variables can be taken into consideration, (6) Mechanistic models depend on available data and are therefore not universal, (7) Neural networks models - These models are emerging and are very promising.

2 ARTIFICIAL IMMUNE SYSTEMS

A biological system, such as that of a human being can be regarded as the ultimate information processing system, expected to provide various ideas to the infrastructure engineering field. Information processing systems can be classified into the following forms (Figure 1):

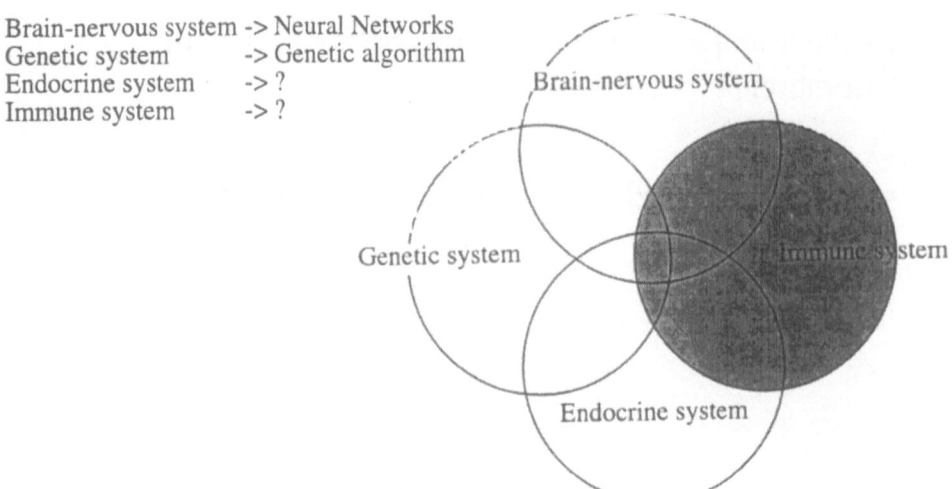

Brain-nervous system -> Neural Networks
Genetic system   -> Genetic algorithm
Endocrine system   -> ?
Immune system   -> ?

Fig. 1. Information processing systems in a biological system

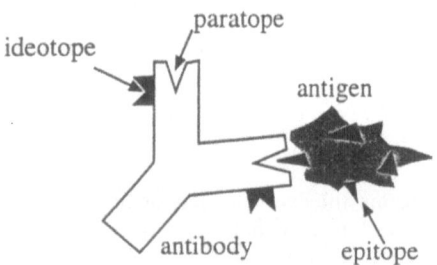

Fig. 2. Structure of an antigen and an antibody

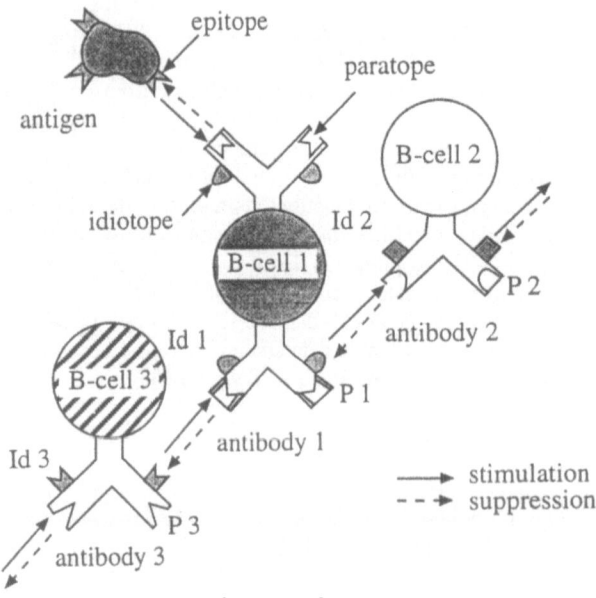

Fig. 3. Jerne's idiotypic network

(1) brain-nervous systems (2) genetic systems,(3) immune systems, and (4) endocrine systems (Ishigoru *et al.* 1994). Among the biological systems, brain nervous and genetic systems have already been applied to infrastructure problems as artificial neural networks (Attoh-Okine 1994) and genetic algorithms (Fwa 1994), respectively. A hybrid system of brain nervous and genetic systems have also been applied to infrastructure systems decision making. However the other systems (endocrine and immune systems) have not been applied to infrastructure systems notwithstanding their important roles in the decision making process.

Immune systems (Dasgupta & Attoh-Okine 1997) have various interesting features such as immunological tolerance, pattern recognition, non-hierarchical distributed systems viewed from an infrastructure standpoint (Ishigoru *et al.* 1995). The artificial immune system (AIS) implements learning techniques inspired by the human immune system which is a remarkable natural defense that learns about foreign substances. The immune system is a rich source of theories and acts as an inspiration for computer-based solutions. Areas of interest relating to the immune network are (Hunt & Cooke 1996): (1) The immune system is a naturally occurring event response system which can quickly adapt to a changing environment, (2) The immune system possesses a self-memory which is dynamically maintained and which allows items of information to be forgotten. It is thus adaptive to its external environment. Immune systems, like neural systems, have a highly sophisticated capability of pattern recognition. However, their recognition and learning mechanisms are quite different from those of a neural system. Natural immune systems are capable of distinguishing virtually any foreign cell or molecule from the body's own cells: known as self and non-self discrimination (Dasgupta & Forrest 1996). The immune system protects our bodies from attaching to foreign substances (called antigens) which enter the blood stream. The basic components of the immune system are lymphocytes that are mainly classified into two types: B-lympocytes and T-lympocytes.

B-lympocytes are the cells produced from bone marrow. B-lymphocytes are contained in the human body, each of which has a distinct molecular structure and produces "Y" shaped antibodies from its surfaces. The antibody recognizes its specific antigens, which are foreign substances that invade living creatures. This reaction is often likened to a lock and key relationship (Figure 2).

T-lymphocytes are the cells produced from the thymus gland, and they generally regulate the production of antibodies from B-lymphocyte. The portion on the antigen recognized by the antibody is called an epitope (antigen determinant) and the one on the antibody that recognizes the corresponding antigen determinant is called a paratope. Furthermore, each type of antibody has its specific antigen determinant called idiotiope. The immune system possesses two types of responses: primary and secondary. A primary response occurs when the immune system encounters the antigen for the first time and reacts against it. The immune system learns about the antigen. The secondary response occurs when the same antigen is encountered again. The secondary response can be elicited from an antigen which is similar, although not identical, to the original one which established the memory.

3 IMMUNE NETWORK MODEL

Jerne (1974) initiated the theoretical development of idiotypic networks which present the mathematical framework of immune systems. Jerne's (1974) idiotypic hypothesis networks are based on the concept that lympocytes are not just isolated, but are communicated to each other among different species of lymphocytes through interaction among antibodies (Figure 3). Jerne initiated the theoretical development of idiotypic networks which presents the mathematical framework of an immune system. His theory is modeled with a differential equation which simulates the dynamics of lymphocytes-: the increase or decrease of the concentration of a set of lymphocyte clones and the corresponding immunoglobins. The idiotypic network hypothesis is based on the concept that lymphocytes are not isolated, but communicate with each other among different species of lymphocytes through interaction among antibodies. Accordingly, the identification of antigens is not done by a single recognizing set but rather a system level recognition of the sets connected by antigen-antibody reactions as a network.

The key postulate of Jerne's theoretical framework is that one cell makes only one antibody. The law, one cell makes one antibody, leads to several predictions which include (1) allelic exclusion, (2) all antibody-like receptors displayed by a lymphocyte should be identical or at least have identical light chains and identical heavy variable regions, and (3) all antibodies produced by a single cell and its progeny should have the identical idiotype.

In formulating the framework, Jerne (1974) discusses formal and functional networks. The formal network discusses repertoires, dualism and suppression. In the discussion of functional networks, a quantitative picture of the theory is presented. According to the functional network, even the absence of antigens that do not belong to the system, must display an eigen-behavior. This mainly results from paratope-idiotope interaction within the system. Jerne concluded that the immune system bears a striking resemblance to the nervous system when viewed as a functional network.

Based on Jerne's work, Perelson (1989), presented a probabilistic approach to idiotypic networks. Perelson's approach is very mathematical, discussing more about phase transition in idiotype networks. Perelson divided phase transition in idiotopic networks to the pre-critical region, transition region and post-critical region. The simulation and suppression chains among lympocytes form a large-scaled network and work as self and non-self recognizers, which provide a new parallel distributed processing architecture unlike neural networks (Ishisgoru, 1996). For example, the idiotope Id1 of antibody 1 stimulates the B- lymphocytes 2, which attached the antibody to its surface through the paratope P2.

Although it appears that the systems are similar to neural network and classifier systems, it differs from these in a number of significant aspects. For example (Hunt & Cook 1996): (1) It is possible to over-teach a neural network such that it does not generalize and can only deal with specific examples. In contrast, artificial immune systems inherently generalize,(2) Learning classifier systems find it difficult to deal with problems which lack separation between global solutions and have many locally optimal solutions. Artificial immune systems can handle these very easily,(3) Artificial neural networks can be very time consuming and tedious to run for a particular application; the artificial immune system is self-organizing,(4) The learning process is achieved in two phases for artificial immune systems: learning of self, which has been the network construction phase, and learning of nonself, whenever new data is presented,(5) The artificial immune systems memory is content addressable, thus is tolerant of noise data,(6) Recognition is achieved not in an active but in a passive manner. That is, rather than identifying the known pattern, it tries to identify nonself whenever there is not a known pattern.

4   NEGATIVE SELECTION ALGORITHM

Forrest *et al.* (1994) developed a negative-selection algorithm for change detection based on the principles of self-nonself discrimination (Percus *et al.*1993) in the immune system. This discrimination is achieved in part by T-cells, which have receptors on their surface that can detect foreign proteins (antigens). During the generation of T cells, receptors are made by a pseudo-randomagenetic rearrangement process. Then they undergo a censoring process, called negative selection, in the thymus where T cells that react against self-proteins are destroyed, so only those that do not bind to self-proteins are destroyed. These matured T cells then circulate throughout the body to perform immunological functions to protect against foreign antigens. The negative-selection algorithm works on similar principles, generating detectors randomly, and eliminating the ones that detect self, so that the remaining T-cells can detect any nonself.

---

[1] This is analogous to the way, proteins are broken up by the immune system into smaller subunits, called peptides, to recognize T-cells receptors

[2] For strings of any significant length a perfect match is highly improbable, so a partial matching rule is used which rewards more specific matches (i.e., matches on more bits) over less specific ones. This

This algorithm approach can be summarized as follows:

- Define *self* as a collection S of strings of length l over a finite alphabet, a collection that needs to be protected or monitored. For example, S may be a program, data file (any software), or normal pattern of activity, which is segmented into equal-sized substrings.[1]
- Generate a set R of *detectors,* each of which fails to match any string in S. Instead of exact or perfect matching,[2] the method uses a partial matching rule, in which two strings match if and only if they are identical at least r contiguous positions, where r is a suitable chosen parameter.
- Monitor S for changes by continually matching the detectors in R against S. If any detector ever matches, then a change is known to have occurred, because the detectors are designed to not match any of the original strings in S.

In the original description of the algorithm, candidate detectors are generated randomly and then tested (censored) to see if they match any self string. If a match is found, the candidate is rejected. This process is repeated until any self string. If a match is found, the candidate is rejected. This process is repeated until a desired number of detectors are generated. A probabilistic analysis is used to estimate the numbers of detectors that are required to provide a certain level of reliability. The major limitation of the random generation approach appears to be a computational difficulty of generating valid detectors, which grows exponentially with size of self. Subsequently, a more efficient detector generation algorithm is proposed (Helman & Forrest 1994) which runs in linear time with the size of self. Other methods for generating nonself detectors have also been suggested (D'haeseleer *et al.* 1996) which have varying degrees of computational complexities.

This algorithm relies on three important principles: (1) each copy of the detection algorithm is unique, (2) detection is probabilistic, and (3) a robust system should detect (probabilistically) any foreign activity rather than looking for specific known patterns of changes.

5 APPLICATION TO PAVEMENT DETERIORATION PROCESS

While some pavement deterioration modeling processes have been addressed (McNiel *et al.* 1992), they fall short of realistically modeling the deterioration processes. During the past few years, there has been considerable research and development related to civil infrastructure systems. Appropriate methods and procedures for infrastructure systems management, advances in deterioration sciences, behavior of materials, new technologies for repair, sensing and computer analysis have been developed. These have served to enhance system characterization, operation and construction, information storage, modeling and other management purposes.

Generally, no artificial immune system applications are used in infrastructure assessment and modeling. Other biologically inspired systems like neural networks (Attoh-Okine 1994 & Schwartz 1993) and genetic algorithms have been used.

Research in technology related to infrastructure has drawn on advances in other disciplines including computer science, statistics, operations research, sensor, robotics, and artificial intelligence techniques. Examples of implementation results include the use of expert systems and knowledge-based systems to assist decision makers in infrastructure management systems, and application of robotics to highway pavement maintenance. Another area which has also helped is the advancement of hardware and software for improving and automating design, construction, decision making, and information collection about the state of the systems. While a major success has occurred in the last few years, there are more opportunities that can improve the assessment, modeling and decision making process in infrastructure. Examples of these successes and opportunities include.

---

partial matching rule reflects the fact that the immune system's recognition capabilities need to be fairly specific in order to avoid confusing self molecules with foreign molecules

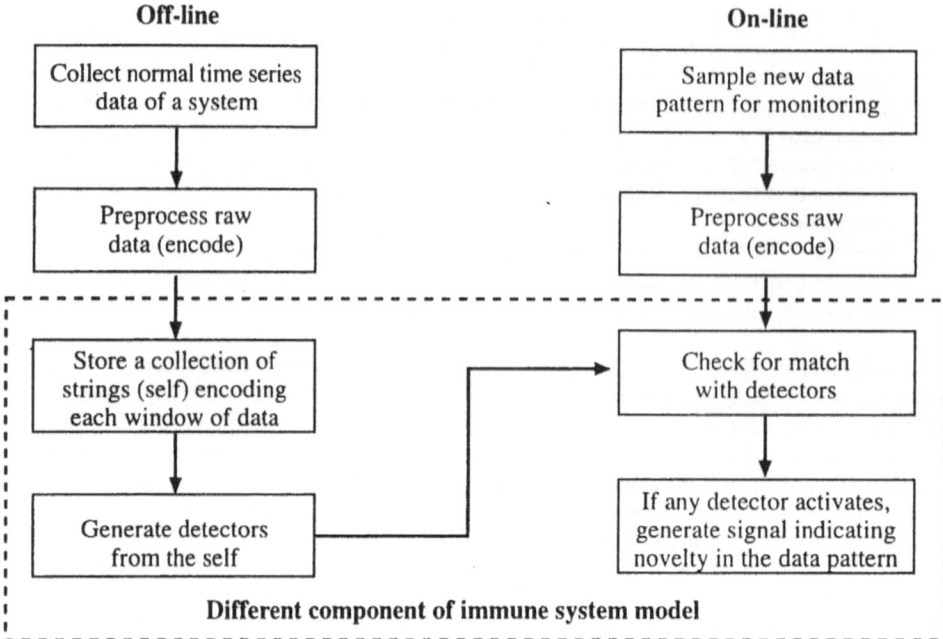

Fig. 4. Schematic diagram showing the processing stages of the immune system based anomaly detection

Dynamically Changing
Infrastructure Environment

Mobile Robot

Infrastructure Condition
System

Normal Infrastructure Behavior   Action Selection

(a) An autonomous mobile robot with an action selection mechanism

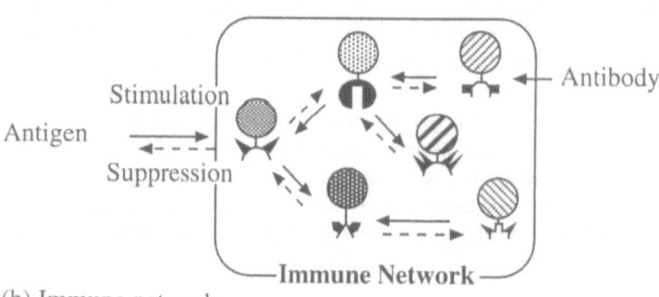

Antigen

Stimulation

Suppression

Antibody

Immune Network

(b) Immune network

Fig. 5. Basic concept of our proposed method

5.1 Information Technologies

Large amounts of data are being collected because of the advances in computer technology, storing and accessing. But, unfortunately, some of the data is very subjective, and incomplete. Improved methods are needed to address the issue of uncertain incomplete and subjective information pavement database.

5.2 Modeling Pavement Performance

Performance measures that capture the nature of the facility, the functionality and its longevity are required for sound management over the facility's own system life. Some measures have been explored for specific types of infrastructure (including highway infrastructure), but the methods fails to address the life cycle and maintenance process included during the infrastructure functioning phase. This is particularly important in the context and the "real" behavior of the system.

5.3 Robotic Systems

Robotic systems play a potential role in highway infrastructure condition assessment. Relatively few of these have been developed in the area of highway pavement infrastructure, which is very limited in its operation and assessment.

5.4 Condition assessment

Sensors for high speed, consistent data collection mean that comprehensive condition data can be collected and analyzed. The normal behavior of an infrastructure system is often characterized by a series of observations over time. The problems of detecting novelties or anomalies in the condition data can be viewed as finding deviation of a characteristic property in the infrastructure system. Artificial immune systems can be used to analyze the instances of a more general problem distinguishing itself (normal condition data not below a threshold value) from other data (deteriorated data below the threshold). The negative selection algorithm (Forest *et al.* 1996) can be applied. The implementation of the approach is presented below (Figure 4).

5.5 Interpretation of Sensor Data

Sensors are widely used in pavement assessment. They provide large quantities of consistent and reliable data. Despite the advances in signal and image processing and considering that most data interpretation (e.g. radar application in highway pavements) is done manually, we may only use samples of data because of time constraints. Research into automated interpretation has been very domain specific but generic issues exist. For example, an interpretation problem may be characterized by the type of data required. The required data may be binary (sound/unsound) or (presence/absence) -e.g. moisture beneath pavement using radar); it may require locating or dimensioning an object or component.

5.6 Faulty States of Pavement Systems

Faulty states of pavement systems are often detected at the propagated points rather than at failure origin. Therefore to realize the exact fault diagnosis, it is necessary to use and integrate the detected data from sensor readings. An immune network of representation of failure diagnosis in highway infrastructure will be presented. In the immune network model, each subsystem is regarded as a distinct antibody. For simplicity, the information about the state of the system can be binary, fault free (normal data) or faulty (abnormal data) and this information can then be treated as an antigen. To realize exact fault diagnoses, variables should be assigned to reflect the state of the system.

5.7 Robotic Systems in Pavement Assessment

Each year the United States spends $20 billion to maintain its roads. The bulk of this money goes to the labor intensive repair of potholes, cracks, and joints of pavement surface. Unfortunately restoring the damaged pavement is rarely permanent. Many workers involved in these labor intensive jobs are tragically killed each year. Emergency repairs are sometimes done in haste, thus minimizing the crew exposure to traffic and decreasing motorist inconvenience. This "throw and go" procedure may result in maintenance activity that last barely than the time it took to make. Ironically, a maintenance crew may fix the same defects and be exposed to traffic hazards several times due to previous hasty repairs.

Haas *et al.* (1992) proposed the application of a robotic system and pavement cracking-sealing system that can be applied as a maintenance activity in pavement infrastructure systems. The development of a prototype immune network-based robotic system equipped with a mobile robot that can be used to select a competence module suitable for detecting the current condition in a bottom up manner. The proposed system can be an immune based action-selection mechanism for highway infrastructure deterioration conditions. Figure 5 schematically illustrates the action-selection mechanism for transportation conditions. In the figure, the current conditions of the infrastructure (cracks-transverse, longitudinal, fatigue) are detected by the installed sensor.

To make the system collect suitable antibodies (maintenance-action) again the current antigen (pavement conditions), the system will function as a condition-action system. For the adequate selection of antibodies, state variables called concentration will be assigned to each antibody (maintenance action). The selection of antibodies (maintenance action) can be simply carried out in a winner-out and winner-take-all fashion, where one antibody is allowed to activate and act its corresponding behavior if its concentration surpasses the pre-specified threshold.

5.8   Hybrid Methods

The Pavement deterioration process can be addressed like an artificial life. The combined use of artificial neural networks, genetic algorithms and immune network systems should be explored. An attempt to combine all these biologically inspired methods leads to powerful computational and scientific interpretation of infrastructure deterioration and decision making. This can be the beginning of the artificial life concept in civil and environmental engineering systems. Rather than building a precise model of immune systems, a new information (mutual recognition and immune network model) will be explored. The mutual recognition network is an immune network information model obtained by focusing the idea of idiotypic network (Ishida & Mizessyn 1992). This can be the beginning of the artificial life concept in civil and environmental engineering systems.

6   CONCLUSION

The artificial immune system (AIS) is a subject of great research interest because of its powerful information processing capabilities. In particular, it performs many complex computations in a completely parallel and distributed fashion. Like the nervous system, the immune system can learn new information, recall previously learned information and performs pattern recognition tasks in a decentralized fashion. Also its learning takes place by an evolutionary processes similar to biological evolution. This paper explores the application of (AIS) in the pavement deterioration modeling process. Using the learning capabilities of AIS, it will be possible to obtain some meaningful insight into deterioration modeling and generally there are many potential application areas in AIS modeling which appear to be useful in civil engineering applications.

REFERENCES

Attoh-Okine, N.O 1994. Predicting Roughness Progression in Flexible Pavements Using Artificial Neural Network. *Proceedings 3rd International Conference on Managing Pavements*. 1: 55-62.

Ben-Akiva, M., Humplick, F., Madanat, S. and Ramaswamy, R. 1993. Infrastructure Management Under Uncertainty: The Latent Performance Approach. *Journal of Transportation Engineering*: 43-57.

Dasgupta, D. & Attoh-Okine, N.O 1997. Immune Based Systems-A Survey. *IEEE International Conference on Systems, Man and Cybernetics*: 369-374. Orlando: Florida.

Dasgupta, D. & Forrest, S 1996. Novelty Detection in Time Series Data Using Ideas from Immunology. Department of Computer, University of New Mexico: Albuquerque.

D'haeseleer, P. 1996. An immunological approach to change detection: theoretical results. *Proceedings of* IEEE *Symposium on Research in Security and Privacy*. Oakland: California.

Fwa, T.F., Chan, W.T. and Tan, C.Y. 1994. Optimal Programming by Genetic Algorithms for Pavement Management. *Paper presented at 1994 Annual Transportation Research Record.*

Feighan, K.J., M. Shahin., Keane, P. and . Wu, M. 1993. Optimal Maintenance Decisions for Pavement Management 3. *Journal Transportation Engineering, ASCE*, 5(113): 554-572.

Forrest, S., Perelson, A. S., Allen, L. and Cherukuri, R. 1994. Self-Nonself Discrimination in a Computer. *Proceedings of IEEE Symposium on Research in Security and Privacy*: 202-212. Oakland : California.

Haas, C., et al 1992. A Field Prototype of a Robotic Pavement Crack Sealing Systems. *Proc. of the 9th International Symposium on Automation and Robotics in Construction.*

Helman Paul & Forrest. Stephanie 1994. An Efficient Algorithm for Generating Random Antibody Strings. *Technical Report No. CS94-7*, Department of Computer Science, University of New Mexico.

Hunt, J.E. & Cooke, D. 1996. Learning Using Artificial Immune Systems. *Journal of Network and Computer Applications*. 19: 189-212.

Ishida, Y. & Mizessyn, F 1992. Learning Algorithm on an Immune Network Model: Application to Sensor Diagnosis. *Proc. of International Joint Conference on Neural Networks*:33-38.

Ishigoru, A., T. Kondo. , Watananbe, Y. and Uchikawa, Y. 1995. Dynamic Behavior Arbitration of Autonomous Mobile Robots Using Artificial Immune Networks. *Proceedings IEEE Conference*: 722-727.

Ishiguro, A., Shirai, Y., Kondo, T and Uchikawa, Y. 1996. Immunoid: An Architecture for behavior Arbitration Based on the Immune Networks, *Paper submitted to IROS 96*. Osaka: Japan.

Ishigoru, A., Watananbe, Y. and Uchikawa, Y. 1994. Fault Diagnosis of Plant Systems Using Immune Networks. *In Proceedings of the 1994 IEEE International Conference on Multisensor Fusion and Intergration for Intelligent Systems (MFI'94)*: 34-42. Las Vegas.

Jerne, N.K. 1974. Toward A Netowork Theory of Immune System, *Annals of Immunology*. 125(C): 373-389.

McNiel, S., Markow, M., Ordway, J. and Uzarki, D. 1992. Emerging Issues in Transportation Facility Management *Journal of Transportation Engineering* 118:4.

Paterson, W.D.O. 1987. Road Deterioration and Maintenance Effects. *Models for Planning and Management*. John Hopkins University Press, Baltimore: Maryland.

Paterson, W.D.O & Attoh-Okine, N.O 1992. Summary Models of Paved Road Deterioration Based on HDM-III. *Transportation Research Record*. 1344, National Academy Press, Washington, DC: 99-107.

Perelson, Alan S. 1989. Immune network theory. *Immunological Reviews* (10): 5-36.

Percus, J. K., Percus, O. and Person, A.S. 1993. Predicting the size of the antibody combining region from consideration of efficient self/ non-self discrimination. *Proceedings of the National Academy of Science*. 60:1691-1695.

Schwartz, C.W. 1993. Infrastructure Condition Forecasting Using Neural Networks. *Proceedings ASCE Conference on Infrastructure Planning and Management*: 282-284.

Artificial Intelligence and Mathematical Methods in Pavement and Geomechanical Systems, Attoh-Okine (ed.)
© *1998 Balkema, Rotterdam, ISBN 90 5809 028 0*

Implications of using higher order stochastic finite element analysis for pavement deflections

Dieter Stolle & Mehdi Parvini
Department of Civil Engineering, McMaster University, Ont., Canada

ABSTRACT: The accuracy of a perturbation based, stochastic analysis depends mainly on the degree of the polynomial expansion used in the formulation. It is understood that for problems involving relatively small variations, a first order analysis provides reasonable predictions of the statistical variations associated with the output. On the other hand, when larger variations of the input variables are anticipated, the first order approximation may no longer be suitable. This paper studies the effect of including higher order terms. Stochastic finite element solutions of a sample pavement problem that were obtained from first and second order accurate covariance analyses are compared in this study. The results of the forward analyses confirm that for larger statistical variations in the input parameters, the difference between the predictions of the two approaches is significant. The effect of systematic errors on the results of the inverse solution to the problem is also briefly addressed.

1 INTRODUCTION

Statistically based techniques of data analysis have been introduced in the civil engineering practice with the objective of developing a more rational approach to the design and evaluation of constructed facilities. The perturbation method is the most widely used technique in practice (Ghanem & Spanos 1991). The attractiveness of this method is its mathematical simplicity. The perturbation scheme consists of expanding all the random quantities around their respective mean values via a Taylor series (Benjamin & Cornell 1970) to obtain a relation between characteristics of the random response and the random structural parameters.

For the practical application of statistical methods to structural analysis, such techniques should be combined with efficient and powerful computational methods. The stochastic finite element method (SFEM) provides a computational tool which models the complexity associated with large engineering systems and takes into account the uncertainties that accompany the random variations in material properties, geometry and loading. Most often analyses are completed using only statistical moments up to the second order for the expected value and first order for the covariance due to the difficulties of finding the higher order moments (Nakagiri & Hisada 1982). This study examines the use of fourth- and second-order approximations for the expected value and covariance, respectively. Owing to the assumption that the random variables are normally distributed, the system characteristics need only be defined in terms of the mean values and the covariances.

2 TAYLOR'S EXPANSION METHOD

The implementation of the Taylor's expansion method in the stochastic analysis is presented in the following sections for scalar and matrix equations.

2.1   Scalar Equation, Low Order Approach

The Taylor's expansion of a function $u(b)$ with respect to its variable b around b_0 is defined by

$$u(b) = u(b_0) + u'(b)\big|_{b_0} \Delta b + \frac{u''(b)}{2!}\bigg|_{b_0} \Delta b^2 + \cdots \tag{1}$$

where $\Delta b = b - b_0$. The prime in Eq. (1) represents differentiation with respect to b, and the notation "$|_{b_0}$" indicates that the quantity is evaluated at b_0. One may consider u to be the displacement of a system and b one of the system properties. For a small Δb, the expansion can be truncated after a few terms which provides an approximation to $u(b)$. Assuming that b is a random variable with b_0 being its mean value, neglecting terms higher than second order, and applying the expected value function to both sides of the truncated expansion (Kleiber & Hien 1992), one has for the expected value of u,

$$E[u(b)] = u(b_0) + \frac{u''(b)}{2!}\bigg|_{b_0} \text{var}(b) \tag{2}$$

in which $\text{var}(b)$ is the variance of random variable b.

To find the variance of function $u(b)$, the first order approximation of $u(b)$ is considered. By taking $u(b_0)$ to the left-hand side of the equation, squaring both sides, and applying the expected value function, the variance is given by

$$\text{var}(u(b)) = [u'(b)\big|_{b_0}]^2 \text{var}(b) . \tag{3}$$

It should be noted that implicit in the calculation of $\text{var}(u(b))$ is the assumption that the expected value $E[u]$ can be approximated by $u(b_0)$.

For the case of a multi-variable function, $u(b_1,\dots b_n)$, Eq.'s (2) and (3) become

$$E[u] = u(b_0) + \frac{1}{2}\sum_{i=1}^{n}\sum_{j=1}^{n}(u^{'b_i'b_j})\big|_{b_{0_i},b_{0_j}} \text{cov}(b_i,b_j) \tag{4}$$

and

$$\text{var}(u) = \sum_{i=1}^{n}\sum_{j=1}^{n}(u^{'b_i})\big|_{b_{0_i}}(u^{'b_j})\big|_{b_{0_j}} \text{cov}(b_i,b_j) \tag{5}$$

with $\text{cov}(b_i , b_j)$ being the covariance of b_i and b_j, and $u^{'b_i}$ denoting the partial differentiation with respect to b_i. In order to simplify the notation, the dependence of a function to its variables is not shown.

2.2   Introduction of Higher Order Terms

If the random variations are not small, the truncated Taylor's expansion should include higher order terms in order to provide reasonable accuracy. To illustrate the effect of higher order approximations, the terms up to the fourth order are retained in the Taylor's expansion of u, yielding,

$$E[u] = u_0 + \frac{u''}{2!}\text{var}(b) + \frac{u'''}{3!}M^3(b) + \frac{u''''}{4!}M^4(b) \tag{6}$$

where $M^3(b)$ and $M^4(b)$ are the third and fourth order central moments of the random variable b, respectively. If the probability density function of b is symmetric, the third order central moment will be zero; i.e., $M^3(b)=0$. Furthermore, given that b is normally distributed, $M^4(b)=3var(b)^2$ (Taylor 1982) and Eq. (6) becomes

$$E[u] = u_o + \frac{u''}{2}var(b) + \frac{u''''}{8}var(b)^2 \tag{7}$$

Following the procedure outlined previously in which u_0 is used as an approximation for $E[u]$, the higher order approximation for var(b) is

$$var(u) = (u')^2 var(b) + \left[0.75(u'')^2 + u'u'''\right]var(b)^2. \tag{8}$$

2.3 *Matrix Equation*

When dealing with a matrix representation of a function, direct differentiation may be cumbersome, thus to derive the equations for the expected value and variance, the following approach is taken. Consider a matrix equation of the form

$$\mathbf{K}(b)\mathbf{u}(b) = \mathbf{Q}(b) \tag{9}$$

which for our purpose represents the equilibrium equation of a multi-degree of freedom system where $\mathbf{K}(b)$ is the stiffness matrix, $\mathbf{Q}(b)$ is the load vector, $\mathbf{u}(b)$ is the displacement vector, and b is a system parameter such as the elastic modulus. Since the stiffness and, in general, the loads are assumed to be functions of the system parameter b, the displacement will also be a function of b.

Given that b is a random variable, it may be expressed in terms of its mean value b_0 and a random perturbation. Assuming Δb to be small and approximating $\mathbf{K}$, $\mathbf{u}$, and $\mathbf{Q}$ with a second-order, truncated Taylor's expansion about b_0, one gets (Kleiber and Hien 1992)

$$(\mathbf{K}_0 + \mathbf{K}'\Delta b + 0.5\mathbf{K}''\Delta b^2)(\mathbf{u}_0 + \mathbf{u}'\Delta b + 0.5\mathbf{u}''\Delta b^2) = \mathbf{Q}_0 + \mathbf{Q}'\Delta b + 0.5\mathbf{Q}''\Delta b^2 \tag{10}$$

where the subscript zero denotes that the quantity is evaluated at b_0, and the prime implies that a function is differentiated with respect to b and evaluated at b_0. Again, to avoid the notation from becoming too complex, the dependence of $\mathbf{K}$, $\mathbf{u}$, and $\mathbf{Q}$ to b is not mentioned.

Expanding the left-hand side of the equation, and collecting terms of the same order gives a set of three equations from which the values of $\mathbf{u}_0$, $\mathbf{u}'$, and $\mathbf{u}''$ are obtained as follows

$$\mathbf{u}_0 = \mathbf{K}_0^{-1}\mathbf{Q}_0 \tag{11}$$

$$\mathbf{u}' = \mathbf{K}_0^{-1}[\mathbf{Q}' - \mathbf{K}'\mathbf{u}_0] \tag{12}$$

$$\mathbf{u}'' = \mathbf{K}_0^{-1}[\mathbf{Q}'' - \mathbf{K}''\mathbf{u}_0 - 2\mathbf{K}'\mathbf{u}']. \tag{13}$$

Considering the truncated Taylor's expansion of $\mathbf{u}$, and following the procedure outlined previously, the expected value vector and the covariance matrix of displacement are given by

$$\mathbf{E}[\mathbf{u}] = \mathbf{u}_0 + 0.5\mathbf{u}''var(b) \tag{14}$$

and

$$\mathbf{cov}(\mathbf{u}) = (\mathbf{u}')var(b)(\mathbf{u}')^t \tag{15}$$

in which $(\mathbf{u'})^t$ is the transpose of $\mathbf{u'}$. The expected value $E[\mathbf{u}]$ is approximated by $\mathbf{u_0}$ when deriving Eq.(15).

For the case of a set of p random variables, and a deterministic load, Eq.'s (14) and (15) change to (Kleiber and Hien 1992)

$$E[\mathbf{u}] = \mathbf{u}_0 + \tilde{\mathbf{u}} \quad , \qquad \tilde{\mathbf{u}} = \sum_{i=1}^{p}\sum_{j=1}^{p} \mathbf{S}_{ij}\mathrm{cov}(b_i, b_j) \tag{16}$$

and

$$\mathrm{cov}(\mathbf{u}) = \mathbf{A}\,\mathrm{cov}(\mathbf{b})\,\mathbf{A}^t \tag{17}$$

in which

$$\mathbf{S}_{ij} = (\mathbf{K}_0^{-1}\mathbf{K}^{'b_i})(\mathbf{K}_0^{-1}\mathbf{K}^{'b_j})\mathbf{u}_0 \qquad i,j=1,...p \tag{18}$$

and $\mathbf{cov(b)}$ is the covariance matrix of the set of random variables $\mathbf{b}$, with $\mathrm{cov}(b_i, b_j)$ representing a member of this matrix, and $\mathbf{cov(u)}$ being the covariance matrix of the random displacements. The superscript b_i beside a prime refers to partial differentiation with respect to b_i. $\mathbf{A}$ is an $n \times p$ matrix consisting of p column vectors defined as

$$\mathbf{a}_i = \mathbf{K}_0^{-1}\mathbf{K}^{'b_i}\mathbf{u}_0 \qquad i=1,...p \tag{19}$$

with $\mathbf{A}^t$ being its transpose. The parameter n represents the number of degrees of freedom.

For the case of the higher order approximation, Eq.'s (16) and (17) are

$$E[\mathbf{u}] = \mathbf{u}_0 + \tilde{\mathbf{u}} + 3\sum_{i=1}^{p}\sum_{j=1}^{p}\mathbf{T}_{ij}\mathrm{cov}(b_i, b_j) \tag{20}$$

and

$$\mathrm{cov}(\mathbf{u}) = \mathbf{A}\mathrm{cov}(\mathbf{b})\mathbf{A}^t + 3\tilde{\mathbf{u}}\tilde{\mathbf{u}}^t + 6\mathbf{A}\mathrm{cov}(\mathbf{b})\mathbf{C}^t \tag{21}$$

in which

$$\mathbf{T}_{ij} = (\mathbf{K}_0^{-1}\mathbf{K}^{'b_i})(\mathbf{K}_0^{-1}\mathbf{K}^{'b_j})\mathbf{S}_{ij}\mathrm{cov}(b_i, b_j) \qquad i,j=1,...p \tag{22}$$

with $\mathbf{C}$ being an $n \times p$ matrix consisting of p column vectors defined as

$$\mathbf{c}_i = \mathbf{K}_0^{-1}\mathbf{K}^{'b_i}\tilde{\mathbf{u}} \qquad i=1,...p \tag{23}$$

3   STOCHASTIC FINITE ELEMENT METHOD

To perform elastostatic, stochastic finite element analyses, the Taylor's expansion approach was incorporated into a finite element program. The code was validated by comparing the results of the stochastic finite element simulations with closed-form solutions for some sample problems (Parvini, 1997).

It should be pointed out that the discretization of a problem for an SFEM procedure is exactly the same as that of any ordinary finite element method. The main difference lies in having to describe material and/or geometric properties in terms of the stochastic variables. The random variables in this paper are described by the expected value and the corresponding variance. When there is more than one random variable, the covariances of the variables, which show their correlations, must also be provided as input, usually by introducing the covariance matrix of the random variables. It should be pointed out that the derivatives of a stiffness matrix with respect to random variables are calculated on the element level, and the resulting matrices are assembled in the usual way, taking into account the compatibility and equilibrium principles. Also, for each random variable it is necessary to have memory at least equivalent to what is required to store the stiffness matrix.

4   SAMPLE PAVEMENT PROBLEM

Figure 1 defines the geometry of a two-layered, pavement-subgrade system consisting of a hot mix surface layer and a subgrade that is supported by bedrock. Material properties are assumed to be linearly elastic and isotropic. The default expected values of the pavement and subgrade elastic modulus are selected to be e_p=4000 MPa and e_s=100 MPa, respectively, with Poisson ratios of v_p=0.35 for the surface layer and v_s=0.45 for the subgrade. The selected values for the pavement properties are consistent with typical resilient (elastic) moduli and Poisson ratios for asphalt concrete reported in the literature (Huang 1993). The subgrade modulus corresponds to the values often encountered when interpreting Ontario FWD data (Stolle 1992). The default thickness of the pavement is assumed to be 0.15 m, with the depth to the bedrock H=7.35 m. A 40 kN load is considered to be applied uniformly over a circular area of radius a=0.15 m.

In order to analyze the problem by the finite element method, the medium is divided into a finite number of elements. Strictly speaking, when dealing with a stochastic analysis, the whole structure should be analyzed using a three-dimensional, finite element code as the variations in layer properties are not necessarily axisymmetric. However, in order to reduce the computational time and to avoid a three-dimensional analysis, it is assumed in this study that the effect of randomness in the horizontal plane is the same for all directions, thereby allowing an axisymmetric formulation. To satisfy the boundary condition on the line of symmetry passing through the center of the loaded area, zero shear conditions are defined along this line. The domain is subdivided into 1200, 4-noded, quadrilateral, isoparametric elements with 30 rows and 40 columns, yielding 1271 nodes and 2430 net degrees of freedom. The mesh is made finer close to the load and coarser as one moves away from the load. The nodes along the bottom boundary are fully fixed, simulating the presence of the bedrock. In order to minimize the effect of an artificial boundary (traction free vertical boundary) on the solution in the region of interest, the domain is extended to 11.85 m away from the centerline, i.e., 79 plate radii away. A schematic of the finite element model is shown in Figure 2. The stochastic analyses were completed using, both, low and higher order approximations.

5   RESULTS OF THE SIMULATIONS

5.1   *Predicted Surface Deflections*

Figure 3 shows the predicted surface deflection profile for the cases where the subgrade modulus is: (a) deterministic (represented by its mean value); and (b) randomly distributed with a coefficient of variation CV(e) = 10% and 25%. Since the deflections are relatively insensitive to the pavement modulus (Parvini, 1997), only the effect of variation in the subgrade modulus on the deflections is illustrated here. A similar exercise, which took into account the variation in the pavement modulus, predicted displacements that were virtually superimposed on those determined from the deterministic analysis. As shown in the figure, a CV of 25% in the subgrade modulus produces a small effect on the deflection basin, with the curve corresponding to 10% having no noticeable effect. Nevertheless, a small systematic error is present when neglecting the natural variation in the subgrade modulus. Although the effect of the variations in the subgrade modulus on the predicted deflection basin is negligible in a forward analysis,

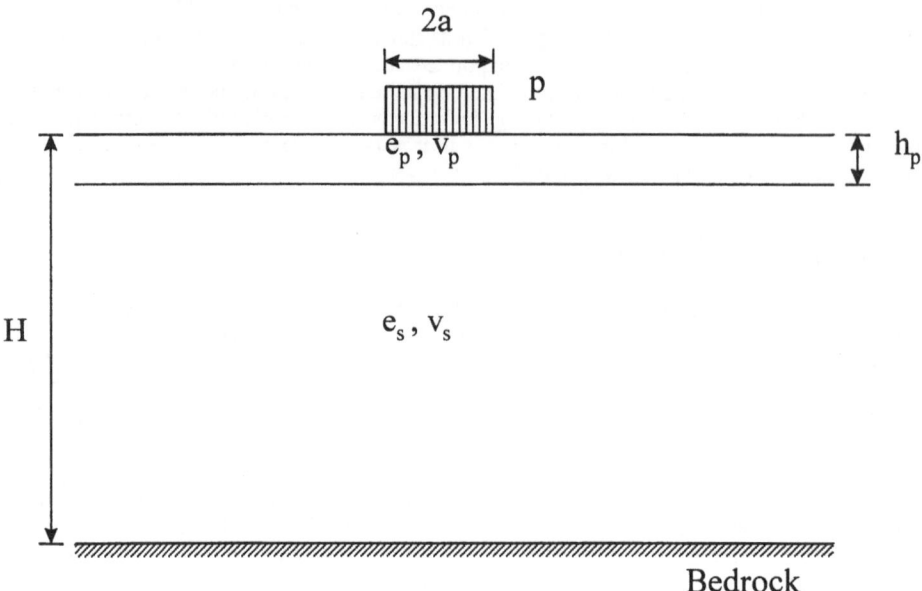

Figure 1   Schematic of a pavement-subgrade system

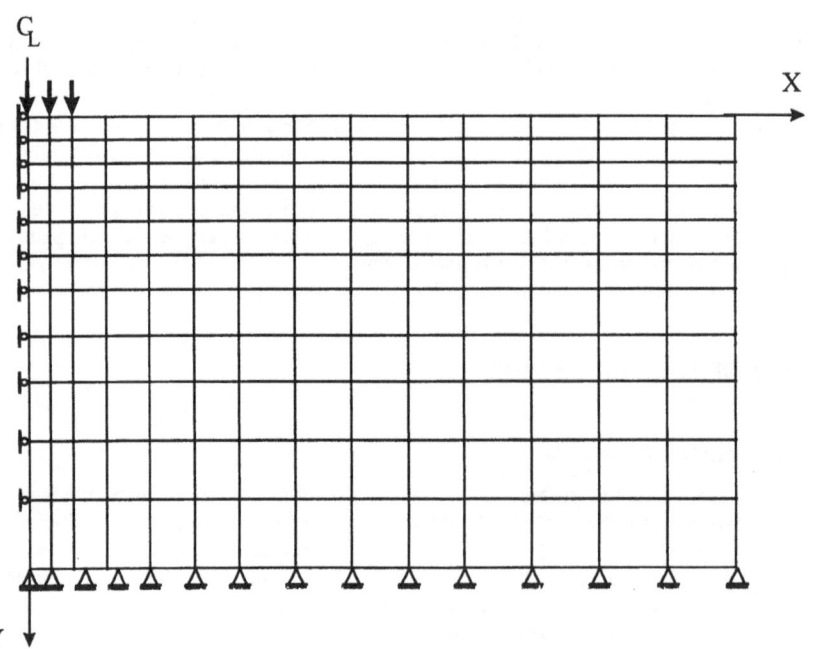

Figure 2   Schematic of the finite element model for a pavement-subgrade system

the variations can be important when solving the inverse problem; i.e., determining a layer modulus given the response due to loading.

As far as the expected value of deflection E[u] is concerned, both low and higher order approximations yield similar results. Although the differences between them may be small, the deflections corresponding to the higher order analysis are greater than those predicted with a low order approximation. The results here confirm that approximating E[u] with u_o, when estimating the variance of deflection, is justified.

5.2   *Variations in Deflections*

Figure 3 provides only the expected values for deflections; i.e., the highly probable values. In actual fact, the deflections may deviate from the expected values. The deviation is described here by the coefficient of variation, which is a dimensionless quantity. The coefficient of variation (Stolle 1992) of deflection, CV(u), is plotted versus the radial offset in Figure 4 for the cases where either the variation in the pavement or subgrade modulus is neglected. The results were obtained using low and higher order approximations together with coefficients of variation of 10 and 25 percent.

As shown in Figure 4, for a specific coefficient of variation of subgrade modulus CV(e), CV(u) increases as one moves away from the center line, reaching a maximum value at approximately 1 m offset. This observation is consistent with the fact that deflections farther from the load are mainly attributed to the straining of the subgrade (Uzan & Lytton 1989). For the first order perturbation method, CV(u) is approximately a linear function of CV(e). On the other hand, the increase in CV(u) is not linear with respect to CV(e) for the higher order analysis. It is clearly shown that, although the random variations in the subgrade modulus do not have much influence on the expected values of deflections, they do affect CV(u). As expected, the difference between the predictions from the low and higher order analyses is negligible when CV(e)=10% while it is significant for the case of CV(e)=25%. The observation clearly demonstrates that the first order analysis is not adequate when high variations exist in the subgrade.

For the case in which the variations are assumed to exist only in the surface layer modulus, the difference between the CV(u)'s predicted by the low and higher order analyses is small when compared to the previous case. Moreover, the difference is negligible for the offsets located farther away from the load. Once again, this is due to the fact that the deflections at larger offsets are related directly to the straining within the subgrade.

6   SOLUTION TO THE INVERSE PROBLEM

The relation between the coefficient of variation of the response CV(u) and that of a layer modulus CV(e) can be interpreted in two ways. Up to this point, the emphasis has been placed on considering CV(u) to provide a measure of the deviation of the response from the expected behaviour. One observes that there is more deviation in the surface deflections than in the subgrade modulus but less deviation than in the pavement modulus. The fact that deviations exist when the modulus of the layers have random variations, indicates that such systematic errors may play a role in the quality of a layer modulus that is backcalculated from a measured response to loading.

The second interpretation is given within the context of sensitivity analysis. It can be shown, at least for the first-order analysis, that the ratio d = CV(u)/CV(e) is a measure of certainty with respect to a backcalculated modulus (Parvini 1997). For d > 1, there is more certainty in the modulus than in the deflections, with the reverse being true for d < 1. Based on deflections at larger offsets, which are often used to estimate the subgrade modulus, one observes that d $\approx$ 1.2. On the other hand, based on the characteristics at a zero offset, which are often used for the pavement modulus predictions, d $\approx$ 0.36. In other words, the quality of a backcalculated subgrade modulus tends to be much greater than that of the pavement. This observation is consistent with experiences in backcalculating layer properties from non-destructive test data (Stolle, 1992).

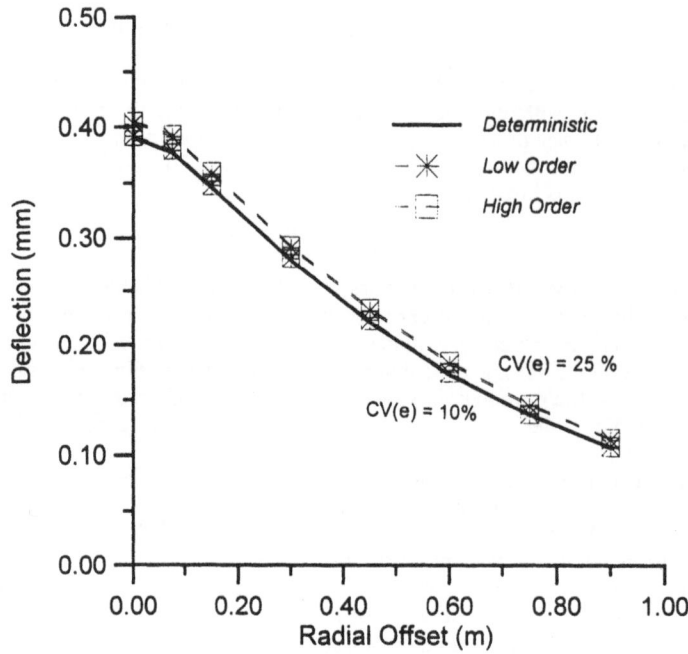

Figure 3   Deterministic and stochastic deflection basins

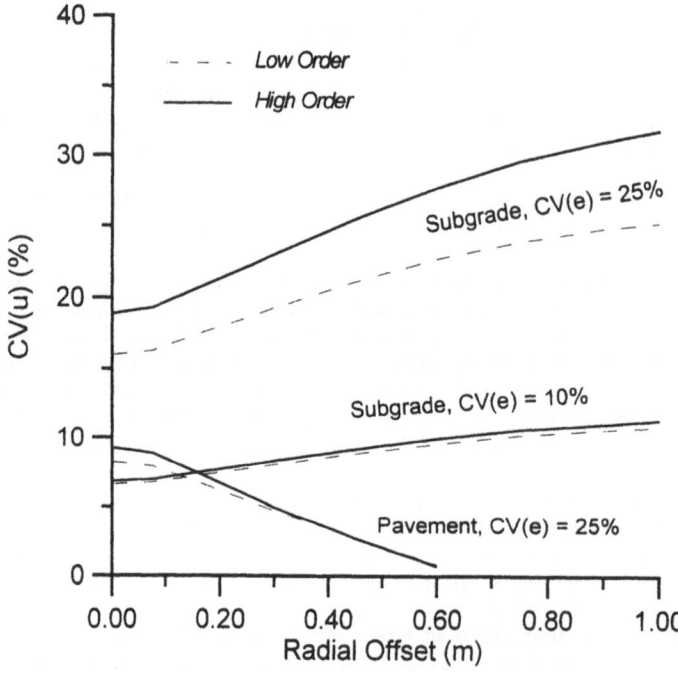

Figure 4   Coefficient of variation of deflection as a function of offset.

7 CONCLUDING REMARKS

An idealized, two-layered, pavement-subgrade system was analyzed within a stochastic finite element framework. It has been demonstrated that systematic errors appear when one neglects the random variations of the stiffness characteristics of a layer. These systematic errors are important in the solution of an inverse problem where the elastic modulus of the pavement and subgrade layers are estimated from the response of the system to loading.

Although the results are not shown in this paper, simulations have been carried out for the other pavement-subgrade configurations. The trends are similar to those illustrated for the sample problem.

ACKNOWLEDGMENTS

The authors wish to thank the Natural Sciences and Engineering Research Council of Canada and the Ministry of Culture and Higher Education of Iran for supporting this research.

REFERENCES·

Benjamin, J.R. & Cornell, C.A. 1970. Probability, Statistics and Decision for Civil Engineers. McGraw-Hill, NY.

Ghanem, R.G., & Spanos, P.D. 1991. Stochastic Finite Element: A Spectral Approach. Springer-Verlag Inc., NY.

Huang, Y.H. 1993. Pavement Analysis and Design. Prentice Hall, Inc., Englewood Cliffs, NJ.

Kleiber, M., & Hien, T.D. 1992. The Stochastic Finite Element Method: Basic Perturbation Technique and Computer Implementation. John Wiley and Sons Inc., Chichester.

Nakagiri, S., & Hisada, T. 1982. Stochastic Finite Element Method Applied to Structural Analysis with Uncertain Parameters. Proc. Int. Conference on the Finite Element Method, pp. 206-211.

Parvini, M., 1997. Pavement Deflection Analysis Using Stochastic Finite Element Method. Ph.D. thesis, McMaster University, Hamilton, Ontario.

Stolle, D.F.E. 1992. Analysis and Interpretation of Falling Weight Deflectometer Data. Ontario Ministry of Transportation. Project No. 21230, Canada.

Taylor, J. R. 1982. An Introduction to Error Analysis, Oxford University Press, CA.

Uzan, J, & Lytton, R.L 1989. Experiment Design Approach to Nondestructive Testing of Pavements. Journal of Transportation Engineering, ASCE, 115(5), pp. 505-520.

Artificial Intelligence and Mathematical Methods in Pavement and Geomechanical Systems, Attoh-Okine (ed.)
© 1998 Balkema, Rotterdam, ISBN 90 5809 028 0

Data mining and analysis for landslide risk using neural networks

T. Kobayashi, H. Furuta, M. Hirokane & S. Tanaka
Faculty of Informatics, Kansai University, Osaka, Japan

I. Tatekawa
Pacific Consultants Inc., Osaka, Japan

ABSTRACT: In this paper we discuss the system for the conclusion of landslide risk by Neural Networks. In this system, we set the parameters to the mesh regions which correspond to the city map, and train this system by the past case of the landslide. Then we give the parameters of the mesh region to this system, and estimate whether the landslide occur or not in the mesh region

1 INTRODUCTION

Landslide is dangerous for people life and it is difficult to estimate the risk that whether it occurs or not. Recently, GIS (Geographical Information System) was developed, we can get the geographical information more easy. Their information has geographical data, such as degree of gradient, we consider that landslide depends on geographical various factors. We already have the past case of landslide, geographical information of each case has become to clear.

In this paper, we will develop the estimation system of landslide risk using neural network trained by past landslide cases, and will examine this system using unknown geographical information.

In the second section, we define the city map, like GIS. In the third section, we describe the structure of neural network. In the fourth section, we have the description of training neural network, in the fifth section, we actually estimate the landslide risk using this system. In the last section, we discuss some problems for this system.

2 DEFINITION OF CITY MAP

In this section, we will define the city map. We show example of city map in Figure 1 The area filled with black means local area. Because this map is example of split meshes, number of meshes becomes less than actual city map.

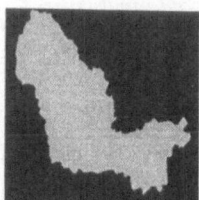

Figure 1: City Map

Each mesh has some attributes, which are 1) existence of the cone, 2) degree of gradient, 3) direction of gradient, 4) distribution of texture, 5) existence of slide, 6) gradient of ground and 7) direction of ground. When put these parameters of one mesh into this system, system provide the grade of landslide risk for the mesh. The system output is 3 parameters which present landslide risk. The landslide risk is presented by 0, 1 and 2. 0 means that landslide has never occurred, 1 and 2 shows the grade that landslide has ever occurred. Thus, the number of input which is represented in following section is 7 and the number of output is 3

We take into consideration Iino city as the example. This city map was split into 380*380 (= 144,400) meshes. In these meshes, the part of the local area is about 54,000 meshes. Further, we split the local area into two parts, left and right part. We use the left part (37,506 meshes), for training this system, and right part (16,504 meshes), for examination of the estimation at beginning. In the following section, according to condition and result of training, these values was descended. Using these descended values, we will train another system.

3   STRUCTURE OF NEURAL NETWORK AND LEARNING METHOD

In this system, we use neural network to estimate the landslide risk. In this section, we describe the structure of neural network in detail.

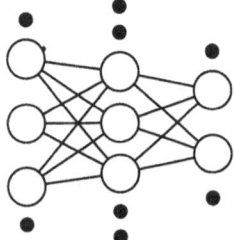

Figure 2: Structure of Neural Network

This network is 3-layered neural network, which have input, hidden and output layer. The unit number of each layer is 7 units, which means number of attributes of each meshes, to input layer, 21 units to hidden layer, 3 units, which means grades of landslide risk, to output layer. The units in the hidden and output layer have the connections between itself and all the units of previous layer.

The method to train this neural network is Back-Propagation Algorithm. The output function f_i of each unit which belong to hidden and output layer is:

$$net_i = \sum_{j=1}^{J} x_j w_{ji} - \theta_j$$

$$f_i = \frac{1}{1 + \exp(-net_i)}$$

where, *i is the unit number, and j is the number of unit for previous layer. About input layer, f_i is:*

$$f_i = x_i^p$$

where, *p* is the number of pattern, x_i^p means the value of ith unit in p^{th} pattern, and θ_i means the threshold value of ith unit.

When put the parameters of one mesh into this system, certain output values is provided by neural network. The relation between these parameters and grades of landslide is already

94

defined, and error values occur between these output and right grades. Using these error values, we train this neural network. The error value is defined as following equation:

$$E_i = \frac{1}{2}(f_i - t_i)^2$$

Where t_i means right grade of landslide.

This equation present Mean Square Error (MSE), and the training of neural network decrease this MSE. In order to decrease the MSE, this error value is differentiated by weight value of between layers:

$$\frac{\partial E_i}{\partial w} = \frac{1}{2}\frac{\partial}{\partial w}(f_i - t_i)^2$$

where **w** is coefficient matrix of connection between units. The training of network is run by modifying this coefficient matrix. The modifying coefficient matrix is run according with following equation:

$$\Delta w_l = \alpha \delta f + \eta \Delta w_{l-1}$$

where α is training coefficient, η is inertial coefficient Δw_l is modifying value of current time and Δw_{l-1} is modifying value of previous time. This equation shows that previous training influence current training. Using this modifying value, we modify connection matrix, and run the training of network.

4   TRAINING OF THE NEURAL NETWORK

Next, we train neural network which was defined previous section. In this section, we describe the training of this neural network in detail. We show case number of each grade for training data. Through described in second section, case number of training data is 37,506. In these cases, the mesh with 'never occurred' is 37,204, 'occurred with grade 1' is 220, and 'occurred with grade 2' is 84. We put the parameters of these meshes into neural network, and train it. The condition with when we train neural network is 1) until MSE becomes specified value below, 2) training coefficient is 0.3 and inertial coefficient is 0.9.

Then, we train this neural network with above condition. The training is run on RISC-based workstation, and the result became below.

Table 1 number of cases for training data

Local area	37,506
Never occurred	37,204
Occurred with grade 1	220
Occurred with grade 2	84

Table 2 number right output for training data

Never occurred	37,204
Occurred with grade 1	5
Occurred with grade 2	1

We show process of error in training in Figure 2. At the point of about 1100 times, the error value descends at minimum, but then glows again. Further the error descends toward the specified value. In this training, because we train this network continuously until the error value becomes specified value below, we didn't intercept the training. The training was terminated by determinate iteration. Using the network terminated training, we examined training data. We show result of examination in Table 2. In this table, we show right number which compare the thing offered to train and the thing this network provided. The rate of right answer is 99.21% for training data.

95

Table 3 number of cases for examination data

Local area	16,506
Never occurred	16,458
Occurred with grade 1	7
Occurred with grade 2	39

Table 4 results of decision of landslide grade

Never occurred	16,506
Occurred with grade 1	0
Occurred with grade 2	0

5   ESTIMATION OF LANDSLIDE RISK

Through previous sections, we defined the data structure and the structure of neural network, and trained the network. In this section, we run this system and will try to estimate landslide risk. We already have trained neural network in this system, and we can put the parameters for input into this system. In second section, we show the part of examination data, right part of meshes. In order to examine this system, we put the parameters of these meshes into this system. The meshes of local area is 16,504. We show number of case for each grade in Table 3. The aim of this section is to put the parameters of these meshes into this system and calculate the rate of right output

When put the parameters of one mesh into this system, the system provide the output values in output layer of neural network. 3 output values is provided, each output represent landslide grade shown above table. We compare these values, and we decide the grade that correspond to the output unit with largest value as landslide grade of the mesh.

Then, we show result of decision of landslide grade in Table 4. To compare this table and Table 3, 'never occurred' grade was right completely. The meshes of 'occurred with grade 1' and 'occurred with grade 2' was written off not at all. The rate of right answer is 99.72% for examination data.

Next, we descend case number of training data. In this training data, there are some patterns with same parameter for input. We except these patterns, and this training data become 8,013 patterns. We take some patterns from among these, and create new training data. 'Never occurred' grade is 100 from 7,952, and 'Occurred with grade 1' is 100 from 44 and 'Occurred with grade 2' is 100 from 17. About 'Occurred with grade 1' and 'Occurred with grade 2', we repeat until it will become the specified number. Then, the total number of training data is 300. The instruction data according to training data is also descended. We train neural network using these new training data, and the rate of right answer is 64.70% with training data and 25.61% with 0% estimation data.

Further, we created new training data. We set the number of each grade to 1,000 patterns. Likewise above, we repeat some patterns about grade 1 and 2. Then, we trained neural network using this new training data, and the rate of right answer is 98.04\% with training data and 98.92% with estimation data. The iteration times of training descended from 4,328 of first example to 253.

6   CONCLUSION

In this paper, we constructed the system to estimate landslide risk using neural network. In first example, the neural network terminated training provided inferior estimation results. There was no right estimation about 'occurred with grade 1' and 'grade 2', since within training data, there are some conflict, that there are some patterns which have same parameters.

In second example, we excepted these patterns, and created new training data with same number of patterns, 100 patterns, for each grade. We trained neural network using this new training data, but the rate of right estimation reduced both training data and examination data.

In third example, we created the training data with same number of patterns, 1000 patterns, for each grade. We trained neural network, and this system achieved the coequal rate of right estimation with first example, but we achieved less training iteration rather than first example.

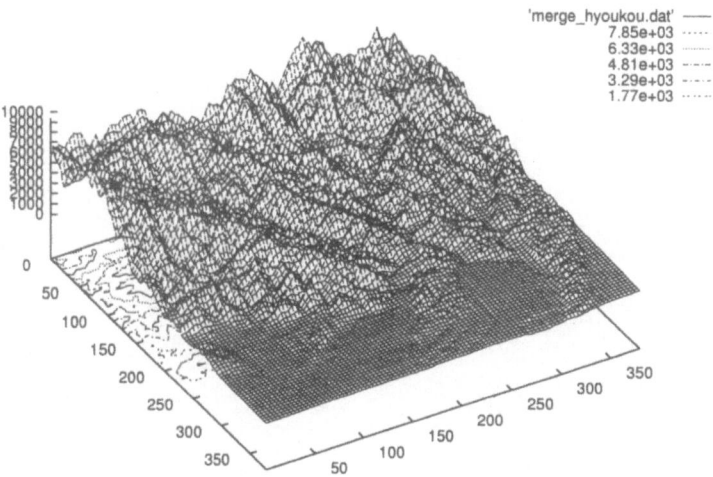

'merge_hyoukou.dat'
7.85e+03
6.33e+03
4.81e+03
3.29e+03
1.77e+03

Figure 3: The bird's-eye view of this city

But, there are still some conflict within the training data, and we think that these conflict influence training and result of estimation. Hereafter we must analyze and dissolve these conflicts.

The second problem is the composition of training data. It is the combination of input parameters and its grade, through above paragraph, there are some conflicts. Now, we create new training data repeating same input and instruction data, but we don't check the conflict about instruction data. We think that there are some patterns that they have same input parameters with different instruction parameters. We must detect these patterns and except either patterns. There are two method for exception of conflict patterns. One method is that we take more grade of landslide in preference to less grade. Another method is backward method of above. Since the aim of this system is estimation of landslide risk, and we think that the first method will be more desirable. Hereafter we must train neural network using this new training data.

The third problem is the selection of training and examination data. The fig.3 shows this city's birds-eye view. The training data set was elected high-difference of elevation region, and the examination data set was elected low-difference of elevation region. We think that it cause this wrong training result and examination result.

REFERENCE:

Arbib, M. A. 1989. *The Metaphorical Brain 2 Neural Networks and Beyond*: John Wiley & Sons Inc. NY
Hertz, J., Krogh, A., Palmer, R. G 1991. *Introduction to the theory of neural computation:* Addison-Wesley Publishing.
Kobayashi, T. *et. al*. 1997. The Estimation of Landslide Risk by Neural Networks. *ECCE Symposium Report*.
Neilsen, R. H. 1990. *Neurocomputing:* Addison-Wesley Publishing.

Artificial Intelligence and Mathematical Methods in Pavement and Geomechanical Systems, Attoh-Okine (ed.)
© *1998 Balkema, Rotterdam, ISBN 90 5809 028 0*

Evaluation of rule-type knowledge extracted by rough set theory from diagnostic cases of slope-failure danger levels

H. Furuta, M. Hirokane, S. Tanaka & T. Kobayashi
Faculty of Informatics, Kansai University, Osaka, Japan

ABSTRACT: Slope-failure is caused by complex working of various factors such as geology, topography, ground, vegetation, rainfall, etc. If the fuzzy theory is applied to all of these factors, results will often be of short reliability. In this paper, we present a method to acquire, removing inconsistencies, the minimum knowledge necessary to deduce reliable results from actual cases where engineers of long experience diagnosed the danger levels of slope-failure. Next comes a method of constructing a knowledge base for an expert system, such acquired knowledge used. Then, described is a method to check the accuracy and reliability of acquired knowledge and evaluate the knowledge base.

1 INTRODUCTION

The failure of slopes caused by heavy rains in the typhoon and *baiu* seasons has taken tolls of lives and done damages all over country (Editorial Department 1972; Takei et al. 1993). To prevent such disaster, it is necessary to take appropriate measures against slope failure at appropriate spots at appropriate times. Preventive works of such structures as are capable of preventing slope failure have to be chosen, slopes surveyed thoroughly, geology, soil properties, costs of civil engineering works, scenery, and so on taken into account (Ishikawa et al. 1995). Slopes requiring such works and their danger levels of failure should be evaluated synthetically, and appropriate spots and times of such works should be determined. Accordingly, it is important to estimate the danger levels of slope failure with accuracy of a certain degree, which also helps to minimize the damages by slope failure.

Slope failure occurs under complex action of various factors, and a number of methods of diagnosing danger levels of slope failure have been proposed; i.e., a method in which a few of the various factors are highlighted (Suzuki et al. 1995), a statistical method (Kobashi, 1995), a dynamic method (Okimura & Ichikawa 1995), and so on. However, in municipal disaster prevention programs, diagnoses of the danger levels of slope failure have been performed based on experiential knowledge of experts, without using any such methods. It can be mentioned as the background of the prevalence of diagnoses based on experiential knowledge of experts that incorporating experts' experiential knowledge in the diagnostic methods proposed to date is difficult and also no established methods are available.

Besides, diagnostic methods of such danger levels have been proposed by agencies and institutions (Ministry of Construction 1983; Forestry Agency, 1982; Expressway Investigation Board, 1977). However, experiential knowledge of experts are required to use any such method, and hence not every civil engineer can use such methods easily.

Under the circumstances, researches into the application of the expert system, (or ES), have been performed in order to establish diagnostic methods on the basis of experiential knowledge of

Table 1 Conditional attributes of slope-failure danger levels

Conditional attributes	Classification of individual situations	Danger ranks
(*a*) Failure spots	1. Large-scale failure spots	a
	2. Many failure spots	b
	3. A few failure spots	c
	4. None	d
(*b*) Signs of failure	1. Signs such as subsidence, crack and displacement	a
	2. None	d
(*c*) Unstable soil mass such as talus	1. Thick	a
	2. Thin	c
	3. None	d
(*d*) Rock heavily weathered or deteriorated	1. Heavy weathering and catchment topography	a
	2. Heavy weathering and deterioration	c
	3. No heavy weathering or deterioration	d
(*e*) Shuttered zone	1. Shuttered zone	b
	2. None	d
(*f*) Gradient of slope	1. Overhanging	a
	2. Over 35	b
	3. 25 35	c
	4. Below 25	d
(*g*) Gully	1. Gully	b
	2. None	d
(*h*) Valley-like depression on slope	1. Outlet of valley-like depression above the road	a
	2. Surface or weathered soil in the depression thicker than surroundings	b
		c
	3. Valley-like depression	d
	4. No valley-like depression	
(*i*) Topography of ground on top of slope	1. Concave (Catchment topography)	b
	2. Flat	c
	3. Convex	d
(*j*) Configuration of cross section of slope	1. Overhanging	a
	2. Flat part in the upper area of the slope	b
	3. Turning point of gradient upward or downward	c
	4. Other than the above	d
(*k*) Water logging	1. Large quantity of water logging	b
	2. Seepage of water logging	c
	3. None	d
(*l*) Cutting of slope for road construction	1. Cutting of thick unstable soil	b
	2. Cutting of bedrock heavily weathered	c
	3. Cutting of bedrock non-weathered	d

experts (Okimura *et al.*, 1992; Nishi, 1992). In these researches, diagnostic systems have been constructed, vagueness, or fuzziness, in experiential knowledge of experts taken into the systems by using the fuzzy theory. However, in applying the fuzzy theory to diagnoses, it is necessary to solve the very difficult and important matter of which factors the theory should be applied to. If the theory is applied to all factors of slope failure and they are included in a diagnostic system, the system may turn out to be of low reliability in diagnosing danger levels (Terano 1987).

The present paper describes a method of applying the rough set theory (Pawlak 1991; Ziarko 1994) to the acquirement of such experiential knowledge. For this purpose, the authors applied the rough set theory to diagnostic cases, by experts, of the danger levels, or decision attributes, of slopes alongside roads and removed conditional attributes and classes of each conditional attribute insignificant in the diagnoses to extract minimal decision algorithm which was still capable of making diagnoses equal to those by experts. The conditional attributes retained in the minimal decision algorithm can be considered the same as the conditional attributes to which the experts attached particular importance in their diagnoses. The significance of the conditional attributes retained in the minimal decision algorithm is evaluated, and a method of deriving rules from the algorithm for the construction of ES's is described.

Table 2 Observation data for diagnosis of slope-failure danger levels

Slopes	Conditional attributes												Danger levels
	(a)	(b)	(c)	(d)	(e)	(f)	(g)	(h)	(i)	(j)	(k)	(l)	
X1	2	1	1	2	2	2	2	4	2	2	3	1	A
X2	4	2	1	2	2	2	2	4	2	2	2	1	B
X3	3	2	2	3	2	2	2	4	3	4	3	2	C
X4	4	2	1	2	1	3	2	2	2	3	3	3	B
X5	3	2	1	1	2	2	2	4	1	3	3	1	A
X6	3	1	1	1	2	2	2	4	3	2	1	1	A
X7	4	2	2	2	2	2	2	4	2	4	3	3	C
X8	4	2	2	2	2	2	2	4	2	4	3	3	C *
X9	3	1	1	2	2	4	2	3	1	4	3	3	B
X10	3	1	1	3	2	4	2	3	1	4	3	3	B
X11	4	2	1	3	2	4	2	4	2	3	3	3	C
X12	4	2	1	3	2	4	2	3	1	3	3	3	B
X13	4	2	1	3	2	2	2	4	2	3	1	1	B
X14	3	2	2	3	2	2	1	4	1	4	2	3	B
X15	4	2	1	3	2	4	1	4	2	4	3	3	C
X16	4	2	1	3	2	2	2	4	3	4	1	1	B
X17	4	1	1	3	2	4	2	3	1	3	2	3	B
X18	3	2	1	3	2	2	1	4	2	3	3	1	B
X19	4	2	1	3	2	2	2	4	2	3	3	1	C
X20	4	2	1	3	2	3	2	3	1	3	3	3	B
X21	3	1	1	3	2	2	2	4	2	3	3	3	B
X22	4	2	2	3	2	2	2	4	3	3	3	3	C
X23	4	2	2	3	2	3	1	4	1	3	3	3	C
X24	4	1	1	2	2	4	2	3	1	4	3	1	B
X25	4	2	2	1	1	2	2	4	1	4	1	1	A
X26	4	2	2	2	2	2	1	3	1	4	1	1	B
X27	4	1	2	3	2	2	2	4	3	4	3	3	C
X28	2	1	1	1	1	2	2	4	2	4	3	1	A
X29	2	1	1	1	1	2	2	4	2	4	3	1	A *
X30	2	1	1	1	2	2	1	4	3	3	3	2	A
X31	3	1	1	2	2	2	1	4	1	3	2	1	A
X32	4	2	2	2	2	2	2	4	3	3	3	2	C

2   DIAGNOSTIC CASES BY EXPERTS

Used in the present study to extract and evaluate experiential knowledge of experts were 32 diagnostic cases of slopes mainly of state alongside roads. These diagnoses of the danger levels were performed by experts in accordance with the diagnostic method prepared by Expressway Investigation Board (Expressway Investigation Board 1977). Removing indistinctive conditional attributes in the diagnostic method, the authors put into Table 1 12 conditional attributes such as (a) failure spots and (b) signs of failure. Each conditional attribute is provided with classes. For example, the conditional attribute of (a) failure spots is provided with 4 classes: 1. large-scale failure spots, 2. many failure spots, 3. a few failure spots, and 4. none. Danger ranks a, b, c and d, are assigned to the classes of each conditional attribute. The danger ranks of all the conditional attributes of each slope are quantified, its total score is calculated, and its global danger level, A, B or C, is determined. Table 2 shows the class Nos. of the conditional attributes and the global danger levels of the 32 slopes diagnosed by the experts. For instance, the slope X1 is classified into the class No. 2 of the conditional attribute (a) and the class No. 1 of the conditional attribute (b), and its danger level is diagnosed as "A." In other words, this table shows the relation between the class Nos. of the conditional attributes of each slope and its danger level, or decision attribute, and such relations and such a table are called "decision rules" and "a decision table," respectively.

Table 3 Decision table after conditional attributes (*h*) and (*i*) are removed

Slopes	Conditional attributes										Danger levels
	(*a*)	(*b*)	(*c*)	(*d*)	(*e*)	(*f*)	(*g*)	(*j*)	(*k*)	(*l*)	
X1	2	1	1	2	2	2	2	2	3	1	A
X2	4	2	1	2	2	2	2	2	2	1	B
X3	3	2	2	3	2	2	2	4	3	2	C
X4	4	2	1	2	1	3	2	3	3	3	B
X5	3	2	1	1	2	2	2	3	3	1	A
X6	3	1	1	1	2	2	2	2	1	1	A
X7	4	2	2	2	2	2	2	4	3	3	C
X9	3	1	1	2	2	4	2	4	3	3	B
X10	3	1	1	3	2	4	2	4	3	3	B
X11	4	2	1	3	2	4	2	3	3	3	C *
X12	4	2	1	3	2	4	2	3	3	3	B *
X13	4	2	1	3	2	2	2	3	1	1	B
X14	3	2	2	3	2	2	1	4	2	3	B
X15	4	2	1	3	2	4	1	4	3	3	C
X16	4	2	1	3	2	2	2	4	1	1	B
X17	4	1	1	3	2	4	2	3	2	3	B
X18	3	2	1	3	2	2	1	3	3	1	B
X19	4	2	1	3	2	2	2	3	3	1	C
X20	4	2	1	3	2	3	2	3	3	3	B
X21	3	1	1	3	2	2	2	3	3	3	B
X22	4	2	2	3	2	2	2	3	3	3	C
X23	4	2	2	3	2	3	1	3	3	3	C
X24	4	1	1	2	2	4	2	4	3	1	B
X25	4	2	2	1	1	2	2	4	1	1	A
X26	4	2	2	2	2	2	1	4	1	1	B
X27	4	1	2	3	2	2	2	4	3	3	C
X28	2	1	1	1	1	2	2	4	3	1	A
X30	2	1	1	1	2	2	1	3	3	2	A
X31	3	1	1	2	2	2	1	3	2	1	A
X32	4	2	2	2	2	2	2	3	3	2	C

Table 4 Combinations of conditional attributes

Cases	Conditional attributes				
Case-1	(*a*)	(*b*)	(*h*)	(*i*)	(*k*)
Case-2	(*a*)	(*c*)	(*d*)	(*h*)	(*k*)
Case-3	(*a*)	(*c*)	(*d*)	(*i*)	(*k*)
Case-4	(*a*)	(*c*)	(*h*)	(*i*)	(*k*)
Case-5	(*a*)	(*d*)	(*h*)	(*i*)	(*k*)
Case-6	(*a*)	(*d*)	(*h*)	(*j*)	(*k*)
Case-7	(*a*)	(*d*)	(*h*)	(*k*)	(*l*)
Case-8	(*a*)	(*h*)	(*i*)	(*j*)	(*k*)
Case-9	(*a*)	(*h*)	(*i*)	(*k*)	(*l*)

3   EXTRACTION OF MINIMAL DECISION ALGORITHM

3.1   Subordination of Danger Levels to Conditional Attributes

First of all, it is necessary to check whether the danger levels were compatible with 12 conditional attributes in Table 2 which shows the summary of the diagnostic results by experts. The decision rules of all the slopes were examined to find non-deterministic rules; i.e., slopes which were classified into one and the same class under every conditional attributes but were

Table 5 Decision table based on the conditional attributes in *Case*-1

Slopes	Conditional attributes					Danger levels
	(a)	(b)	(h)	(i)	(k)	
X1	2	1	4	2	3	A
X2	4	2	4	2	2	B
X3	3	2	4	3	3	C
X4	4	2	2	2	3	B
X5	3	2	4	1	3	A
X6	3	1	4	3	1	A
X7	4	2	4	2	3	C
X9	3	1	3	1	3	B
X12	4	2	3	1	3	B
X13	4	2	4	2	1	B
X14	3	2	4	1	2	B
X16	4	2	4	3	1	B
X17	4	1	3	1	2	B
X18	3	2	4	2	3	B
X21	3	1	4	2	3	B
X22	4	2	4	3	3	C
X23	4	2	4	1	3	C
X24	4	1	3	1	3	B
X25	4	2	4	1	1	A
X26	4	2	3	1	1	B
X27	4	1	4	3	3	C
X30	2	1	4	3	3	A
X31	3	1	4	1	2	A

assigned different danger levels. Non-deterministic rules were not found in Table 2, and hence the danger levels proved subordinative to the conditional attributes. If non-deterministic rules are found in such a decision table, it means that the number of conditional attributes in the decision table is not sufficient and new conditional attributes have to be added to the existing ones. In the process of extracting minimal decision algorithm, it is necessary for the time being to make trial and error to rectify non-deterministic rules, if any, and make a decision table free of contradictions.

On the other hand, the slope X7 and X8 were governed by one and the same rule. This was also true of the slopes X28 and X29. In such a case, it suffices to remove one slope and consider only the other. Accordingly, the slopes X8 and X29 which are marked with "*" were removed from Table 2 to obtain a new decision table.

3.2   Reduction of Conditional Attributes

Next, it was necessary to find conditional attributes insignificant in the diagnoses. A number of conditional attributes were removed each time, and it was checked whether any contradiction occurred or not in the decision table (Table 2 minus the slopes X8 and X29). Table 3 shows an example, where the conditional attributes (h) and (i) are removed, and the decision rules of the slopes X11 and X12 which are marked with "*" are contradictory to each other, which proves that the danger level of each of the slopes X11 and X12 is subordinative to one or both of the conditional attributes (h) and (i), and the two conditional attributes can not be removed simultaneously. Every combination was removed from Table 2 minus the slopes X8 and X29, and it was checked whether any contradiction occurred or not among the decision rules. Table 4 shows the nine combinations of conditional attributes, *Case*-1 to 9, each of which consists of the minimum number of conditional attributes but still is able to diagnose every slope without contradiction. *Case*-1 taken as an example renders a description of how to extract the minimal decision algorithm follows.

The conditional attributes other than those of (a), (b), (h), (i), and (k) of *Case*-1 were removed from Table 2 minus slopes X8 and X29 to obtain a new table. In this table, a pair of the slopes X1 and $X28$, a group of the slopes $X7$, $X11$, $X15$, and $X19$, and so on governed by one and the same rule each, and the slopes of each pair or group were removed except one to obtain Table 5.

3.3   Reduction of Classes of Conditional Attributes

Finally, it was necessary to examine the classes of the conditional attributes in Table 5. The class Nos. of each conditional attribute were removed one by one to see whether any contradiction occurred or not. If the class No. of a conditional attribute of a slope is removed in such a decision table and contradiction occurs, the class No. proves significant in the diagnosis of the slope. If not, it proves insignificant in the diagnosis. For example, when the class No. 2 of the conditional attribute (a) of the slope $X1$ has been removed in Table 5, we can obtain a new decision table. In this table, the slope $X1$ is assigned No. 1, 4, 2, and 3 in the columns of the conditional attributes (b), (h), (i), and (k) and its danger level is diagnosed as "A," while the slope $X21$ is assigned the same Nos. in the columns of the same conditional attributes but its danger level is diagnosed as "B." Thus, contradiction has been brought about between the decision rules of the slopes $X1$ and $X21$ which are marked with "*," proving that the class No. 2 of the conditional attribute (a) of the slope $X1$ is significant in the diagnosis of the slope $X1$ and hence can not be removed.

By removing the class numbers one by one as mentioned above, a new decision table was obtained. Notwithstanding many class Nos. removed, this table contains no contradiction and is capable of making diagnoses equal to those by Table 5. Besides, in this table, a pair of the slopes $X1$ and $X30$, a group of the slopes $X9$, $X12$, $X17$, $X24$, and $X26$, and so on were governed by one and the same rule each, and accordingly the slopes of each pair or group were removed expect one to obtain a new decision table, or Table 6. This table contains no single conditional attribute or class removable without causing contradiction and is called "minimal decision algorithm."

4   EVALUATION OF MINIMAL DECISION ALGORITHM

Figure 1 shows the frequencies of appearance in Table 4 of each of the conditional attributes which are required by the minimal decision algorithm mentioned above. The conditional attribute (a) is required by every case's minimal decision algorithm, its appearance frequency being nine. On the other hand, the conditional attribute (b) is required by Case-1's minimal decision algorithm alone, its appearance frequency being one. If we rank the conditional attributes in the descending order of the appearance frequency, it turns out as follows: (a) failure spots, (k) water logging, (h) valley-like depression on slope, (i) topography of ground on top of slope, and (d) rock heavily weathered or deteriorated.

Table 6 Minimal decision algorithms in Case-1

| Slopes | Conditional attributes | | | | | Danger levels |
	(a)	(b)	(h)	(i)	(k)	
$X1$	2	--	--	--	--	A
$X2$	--	--	--	2	2	B
$X3$	--	2	--	3	3	C
$X4$	--	--	2	--	--	B
$X5$	3	--	4	1	3	A
$X6$	--	1	--	--	1	A
$X7$	4	--	4	--	3	C
$X9$	--	--	3	--	--	B
$X13$	--	--	--	2	1	B
$X14$	--	2	--	--	2	B
$X16$	--	2	--	3	1	B
$X18$	3	--	--	2	--	B
$X25$	--	--	4	1	1	A
$X27$	4	--	--	3	3	C
$X31$	--	1	4	--	2	A

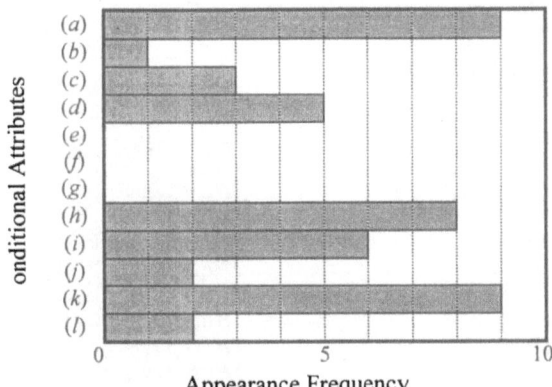

Figure 1 Appearance frequency of conditional attributes

The conditional attribute (*a*) relates to slope histories. Two slopes adjacent to each other can be considered to have similar geology and soil properties, unless there is some extreme change of the geologic structure between the slopes. Accordingly, if there is a failure spot nearby, the slope is most probably regarded to have a high danger level of failure. A spot where minute changes of distances between contour lines continue or a spot where vegetation is different from that of its surroundings is indicative of slope failure (Okuzono 1986). Thus, the conditional attribute (*a*) for failure histories is one of the important conditional attributes in diagnoses of danger levels of slope failure.

The conditional attribute (*k*) relates to water logging. As suggested by the fact that slope failure often occurs under localized heavy rains, water such as rain and drainage is an important conditional attribute of slope failure. Any of the conditional attributes (*h*), (*i*), and (*d*) has a class or classes for catchment topographies of grounds on top of slopes. Known for a long time as a conditional attribute of slope failure is a topography in which water such as rain water and drainage is collected to the area on top of a slope (Okuzono 1986). Thus, the conditional attributes (*k*), (*h*), (*i*), and (*d*) all relating to water paths can be considered important conditional attributes in diagnoses of danger levels of slope failure.

5 DERIVATION OF RULES

The minimal decision algorithm mentioned above can be described easily by rules such as "*IF* conditional part (conditional attribute) *THEN* conclusive part (danger level, or decision attribute)." The minimal decision algorithm of *Case*-1 shown in Table 6 can be described by the 15 rules shown in Table 7. The rules 1 and 2 express the decision rules of the slopes $X1$ and $X2$, respectively. The rule 15 represents the decision rule of the slope $X31$.

For instance, the decision rule of the slope $X1$ in Table 6 indicates that if a slope is classified into the class No. 2 of the conditional attribute (*a*), its danger level is "A." This can be described by a rule below:

IF (*a*) = 2

THEN Danger Level = A

The decision rule of the slope $X2$ indicates that its danger level is "B" when both the conditions are met at the same time, one condition being that a slope is classified into the class No. 2 of the conditional attribute (*i*) and the other being that the slope is classified into the class No. 2 of the conditional attribute (*k*). This can be described by a rule below:

IF (*i*) = 2 and (*k*) = 2

THEN Danger Level = B

Table 7 Rule-type description example of the minimal decision algorithm in *Case*-1

Rule No.	Rule-type description	
1	*IF* (*a*) = 2	*THEN* Danger level = A
2	*IF* (*i*) = 2 *and* (*k*) = 2	*THEN* Danger level = B
3	*IF* (*b*) = 2 *and* (*i*) = 3 *and* (*k*) = 3	*THEN* Danger level = C
4	*IF* (*h*) = 2	*THEN* Danger level = B
5	*IF* (*a*) = 3 *and* (*h*) = 4 *and* (*i*) = 1 *and* (*k*) = 3	*THEN* Danger level = A
6	*IF* (*b*) = 1 *and* (*k*) = 1	*THEN* Danger level = A
7	*IF* (*a*) = 4 *and* (*h*) = 4 *and* (*k*) = 3	*THEN* Danger level = C
8	*IF* (*h*) = 3	*THEN* Danger level = B
9	*IF* (*i*) = 2 *and* (*k*) = 1	*THEN* Danger level = B
10	*IF* (*b*) = 2 *and* (*k*) = 2	*THEN* Danger level = B
11	*IF* (*b*) = 2 *and* (*i*) = 3 *and* (*k*) = 1	*THEN* Danger level = B
12	*IF* (*a*) = 3 *and* (*i*) = 2	*THEN* Danger level = B
13	*IF* (*h*) = 4 *and* (*i*) = 1 *and* (*k*) = 1	*THEN* Danger level = A
14	*IF* (*a*) = 4 *and* (*i*) = 3 *and* (*k*) = 3	*THEN* Danger level = C
15	*IF* (*b*) = 1 *and* (*i*) = 4 *and* (*k*) = 2	*THEN* Danger level = A

Table 8 Total number of rules and conditions to be checked

Cases	Total number of rules	Total number of conditions
Case-1	15	35
Case-2	12	26
Case-3	16	37
Case-4	15	32
Case-5	15	32
Case-6	15	35
Case-7	15	35
Case-8	18	42
Case-9	17	39

To determine danger levels of conclusive parts by using an ES constructed on the basis of the above rules, it is necessary to check whether every condition in each rule is satisfied or not. As for the rule 1, it has to be checked whether the slope is classified into 2 of (*a*) or not. As for the rule 2, it has to be checked whether the slope is classified into 2 and 2 under (*i*) and (*k*), respectively. The total number of conditions to be checked in the 15 rules of *Case*-1 shown in Table 7 is 35. On the other hand, the decision tables of the minimal decision algorithm were also extracted for *Case*-2 to 9, and the numbers of rules and the total numbers of conditions to be checked of all the cases are summarized in Table 8.

Rules for the construction of ES should be so described that their number can be minimized for the sake of knowledge renewal and so on. Besides, conditions to be checked also should be so described that their number can be minimized for the sake of speedy reasoning. Accordingly, efficient renewal of knowledge and speedy reasoning become possible if an ES is constructed on the basis of the rules derived from the minimal decision algorithm of *Case*-2.

6   CONCLUSIONS

The relevant information about the decision rules based on rough set methods was introduced and the application method of the theory to decision-making problems was described. Then, discussed and evaluated was the method of extracting experiential knowledge of expert from the diagnostic results of danger levels of slope failure as the decision tables of minimal decision algorithm. Further discussed was a method to derive rule-type knowledge necessary for the construction of ES from the minimal decision algorithm.

Main findings of the present study are as follows:

(1) Minimal decision algorithm could be extracted from the diagnostic results of slope failure by experts, and conditional attributes significant in the diagnoses were identified. They related to water logging, catchment topographies of grounds on top of slopes, and failure histories of slopes which have been considered important from long ago.

(2) The conditional attributes indispensable for the minimal decision algorithm were the same as those regarded as important from long ago. Accordingly, conditional attributes which we attach particular importance to can be identified by extracting minimal decision algorithm from diagnostic results by an engineer, and his skills can be evaluated.

(3) The decision table stipulates what decision to make when a certain condition is met. Almost all decision-making problems can be formulated by such decision tables.

(4) By removing a conditional attribute or attributes each time in a decision table and seeing whether any contradiction will occur or not, we can evaluate whether the removed conditional attribute or attributes are significant in the diagnoses or not.

(5) Because rough set theory start from the problem of how to express the set of decision attributes by the set of paired conditional attributes and classes, the rough set theory can be used as a method to acquire knowledge from existing diagnostic cases in the field of civil engineering.

REFERENCES

Editorial Department of execution technique: Total of recently torrential rain, Execution Technique, Vol.5, 1972 (in Japanese).

Expressway Investigation Board: Study on investigation method in preventing landslide and slope failure, 1977 (in Japanese).

Forestry Agency: Investigation point in the dangerous area of slope failure, 1982 (in Japanese).

Ishikawa, Y., Furuzeki, J., Sasaki, Y., et al.: Soil attack -- Soil disaster, Geotech note, J. of Japanese Geotechnical Society, 1995 (in Japanese).

Kobashi, S.: Problems in classifying the danger levels of slope failure, J. of Landslide, Vol.10, No.3, 1995 (in Japanese).

Ministry of Construction (Sand Arrestation Section): Manual of countermeasure works for slope failure, Conference of Landslide and Slope Failure, 1983 (in Japanese).

Nishi, K., Furukawa, K., Nakagawa, K.: An evaluation system for slope-failure possibility factors using fuzzy set theory, J. of Japan Society of Civil Engineers, No.445/III-18, 1992 (in Japanese).

Okimura, T., Hirokane, M. et al.: Diagnosis of the slope-failure danger levels based of fuzzy expert system, Kansai Branch of Japan Society of Civil Engineers, 1992 (in Japanese).

Okimura, T., Ichikawa, R.: Prediction method of the slope-failure danger levels based on the numerical topographic features model, J. of Japan Society of Civil Engineers, Vol.358, 1995 (in Japanese).

Okuzono, S.: One hundred point in preventing soil disaster, Association of Kajima pub., 1986 (in Japanese).

Pawlak, Z.: Rough sets -- Theoretical aspects of reasoning about data, Kluwer Academic Pub., 1991.

Suzuki, M. et al.: Dangerous rainfall which soil disaster may be happened, J. of New Erosion Control, Vol.110, 1995 (in Japanese).

Takei, A., Kobashi, S., Nakayama, M., et al.: Landslide, slope failure and debris flow -- Prediction and countermeasures, Association of Kajima Pub., 1993 (in Japanese).

Terano, K., Asai, K., Sugano, M.: Introduction to fuzzy system, Ohmu Pub., 1987 (in Japanese).

Ziarko, W.: Rough sets, fuzzy sets and knowledge discovery, Springer Verlag, 1994.

Artificial Intelligence and Mathematical Methods in Pavement and Geomechanical Systems, Attoh-Okine (ed.)
© *1998 Balkema, Rotterdam, ISBN 90 5809 028 0*

Data warehousing and decision making in construction organizations

I. Ahmad & C. Nunoo
Department of Civil and Environmental Engineering, Florida International University, Miami, Fla., USA

ABSTRACT: In today's global and competitive construction industry, the need to make better and quicker decisions is critical; and the key to success is information. Timely access to useful and meaningful information can enable construction companies gain competitive edge, increase client satisfaction, expand market share and enhance profitability. It is evident that mounting business and economic pressures are forcing companies to radically redesign their information technology infrastructures. Vast amounts of construction operational data are scattered across multiple, dispersed and fragmented departments, units or project sites. To support collaborative construction, it is important not only to share the information, but also to manage that information in a manner that promotes meaningful integration. In this paper, we present data warehousing as an emerging database management technology that can provide the platform for information integration. We point out the difference between an operational database - used for transaction processing; and data warehouse - intended to be used for analytic processing in management decision making in the context of construction organizations. In addition, we show how data warehousing concept can be integrated into decision support systems of a construction organization.

1 INTRODUCTION

Decision makers in the construction industry are under considerable competitive pressure to base their analysis of current business situations on hard facts; facts that are held captive in on-line transaction processing (OLTP) systems and are not readily accessible to them. The concept of data warehousing is fast becoming popular among the decision-makers in many industries particularly in the manufacturing sector (Veshosky 1998). As the construction industry is becoming an increasingly global and multi-organizational sector, involving a large number of participants, working concurrently at different locations and using heterogeneous technologies (Rezgui *et. al.* 1998), the application of data warehousing has a high potential of revolutionizing construction business practices.

As in most organizations, many construction organizations use vast amount of operational data used for transaction processing that are distributed across various functional databases. However, for the most part, the fundamental value of these scattered data is yet to be uncovered. Decision-makers often have to wait days or weeks for responses from IT or MIS personnel that handle requested database queries in order to tap into such data. Such long waiting periods can cause irreversible damages to the reputation of a construction organization Accessing the appropriate information from these databases, at the time when it is needed and in appropriate form, is not an easy task. Data warehousing is presented in this paper as a concept that is meant for gaining timely access to these data in order to facilitate effective decision-making.

According to Inmon H. (1992a), data warehousing is a collection of decision support technologies, aimed at enabling the knowledge worker (executive, manager, and analyst) to make better and faster decisions. Many organizations are committing considerable human, technical, and financial resources to building and using data warehouses. The primary purpose of these efforts is to provide easy access to specially prepared data that can be used with decision support applications, such as management reporting, queries, decision support systems, and executive information systems. Data warehousing is broad in scope, including: extracting data from operational systems and other external sources; cleansing, scrubbing, and preparing data for decision support; maintaining data in appropriate data stores; accessing and analyzing data using a variety of end user tools; and mining data for significant relationships. It obviously involves technical, organizational, and financial considerations.

Data warehousing technologies have already been successfully deployed in many industries. Manufacturing (for order shipment and customer support), retail (for user profiling and inventory management), financial services (for claims analysis, risk analysis, credit card analysis, and fraud detection), transportation (for fleet management), telecommunications (for call analysis and fraud detection), utilities (for power usage analysis), and healthcare (for outcomes analysis). (Adriaans & Zantinge 1996.)

In this paper we will discuss the role of data warehousing, in strategic thinking with special reference to construction organizations. A review of the differences between operational database and that of the data warehouse, general architecture of the data warehouse, online analytic processing (OLAP) used to analyze data extracts from the data warehouse, as opposed to online transaction processing (OLTP) will be presented. Finally, potential benefits that could be derived from its implementation within a construction organization and some institutional issues will be discussed.

2   UNDERSTANDING DATA WAREHOUSING

In general terms a data warehouse is a database created by combining data from multiple databases for the purposes of analysis. They are used solely for reporting purposes unlike the traditional data capture or online-transaction processing (OLTP) systems such as general ledger, accounts payable, financial management, order entry, equipment inventory. Data warehouse is populated with data from two sources. The most frequent source is the periodic migration of data from OLTP systems. The second source is from external purchased databases such as list of incomes and demographic information that can be linked to internal data.

The warehouse gives management the ability to access and analyze information about its business. It allows users to utilize business and market information for directing corporate objectives. The data warehouse collects all of the data into one system, organizes the data so it is consistent and easy to read, keeps "old" data for historical analysis, and makes access to and use of data easy so that users can do it themselves (Corey *et al.* 1998).

From the above general discussion, data warehousing can be defined as a *subject oriented, integrated, non-volatile, time variant* collection of data for management decision (Inmon 1992c). Understanding these key characteristics of data warehouse is crucial for the understanding of how it can be effectively deployed to facilitate efficient and effective decision making system

Subject-oriented data

A data warehouse is established around all the existing applications of the operational data of an organization and it gives information about a particular subject about an organization's on-going operations (e.g., equipment, supplier, orders, shipments) instead of around applications (e.g., general ledger or payroll). Since the data warehouse is designed specifically for decision support while operational database contain information for day-to-day use, not all the information in the operational database is useful for a data warehouse. This makes a data warehouse a flexible information resource for questions that no one has asked before.

Integrated data

In an operational data environment many types of information used in a variety of application may be found and different names for the same entity may be present. For example, only one name must exist to describe each individual entity. In this way the data warehouse reconciles the different data representations in various operational databases used by the organization. To create a useful subject area, the source data must be integrated by modifying it to comply with common coding rules. The data warehouse gathers data into the warehouse from all these variety of sources and merges into coherent whole. In this way they reflect the business information of the organization.

Time-variant data

Most business analyses are, in fact, analyses of trends. Trend analysis requires access to historical data. The data warehouse stores data that can be identified with particular time periods. This implies that there must always be a connection between the information in the data warehouse and the time it was entered. This enables information to be derived according to period. The data warehouse therefore maintains historical data (typically three to five years of data) to enable the knowledge worker to do trend analysis over time. Operational systems on the other hand, are designed for immediate response and therefore data is often purged within a few months of its capture.

Non-volatile data

Data in the data warehouse is never updated but only used for queries. It contains stable data, which are normally read-only environments that are enriched only on a nightly, or weekend basis. In this way such data can only be loaded from the operational database of the organization since that is the only database that can be updated, changed or deleted. Because the data is non-volatile in the data warehouse, it gives management a consistent picture of the state of affairs of an organization's business.

3   DECISION SUPPORT VS. TRANSACTION PROCESSING

The data warehouse is maintained separately from an organization's OLTP databases to minimize the impact that queries have on operational systems and safeguard operational data from being changed or lost. It also allows database administrators to combine fields from different systems to create new, subject-oriented data that end users can access directly using powerful graphical query and reporting tools. This enables users to navigate large corporate data stores in an ad hoc and interactive fashion without disturbing the production data. Another reason for separating the data warehouse from an organization's OLTP databases are that the data warehouse supports on-line analytical processing (OLAP) which enables users to leverage the information stored in databases for sophisticated decision support analysis. The functional and performance requirements of these systems are quite different from those of the on-line transaction processing (OLTP) applications traditionally supported by the operational databases (Zhuge *et al.* 1995).

' The operational database is clearly in the operational environment with data entering into it from unintegrated environment applications. Probably the most important difference between the operational database and the data warehouse is found in the structure and content of data that they are made of. The relationship and difference between the operational database and data warehouse is presented in Figure 1.and Table 1. The two environments and the data each accommodates with the flow of information across the environment are shown in the figure. Once the data has been successfully populated into a well designed data warehouse it can then be made available to all the management decision systems in the organization, namely, Management Information Systems (MIS), Executive Information System (EIS) and Decision Support Systems (DSS). With such major differences in the two environments, technology for supporting them should also be different (Adriaans & Zantinge, 1996).

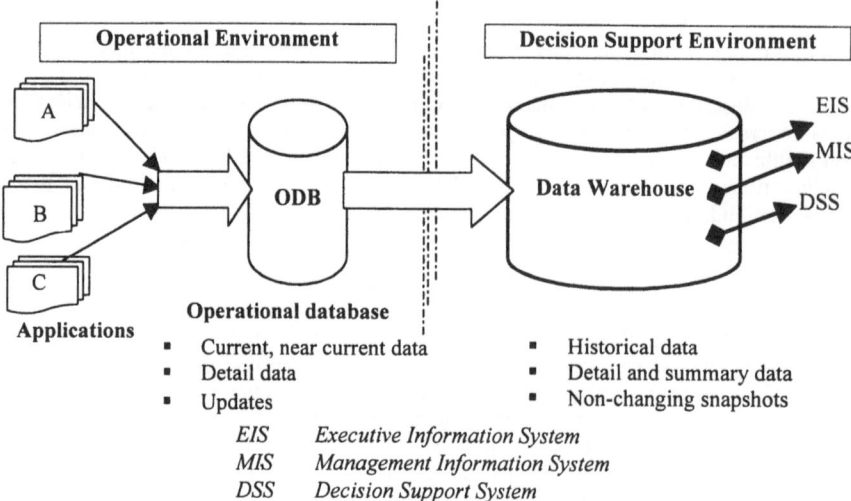

Figure 1: Operation vs. Decision Support Environment (adopted from Adriaans & Zantinge 1996).

While the data warehouse require only basic *load and access* technology with no facility for updates, the operational database needs full-function, general purpose update-record-oriented environment with fast response time.

Since the data warehouse is designed specially for decision support queries, only data that is needed for decision support is extracted from the operational database and stored in the data warehouse. Most online transaction processing (OLTP) environments do not support the four criteria of a data warehouse devised by Bill Inmon we discussed in the previous section. The major differences between decision support (data warehouse) and transaction processing systems are summarized in table 1.

Table 1. Major differences between data warehousing and transaction processing (The Data Warehouse, 1998 & The Role of OLAP Server, 1997)

	Transaction Processing Systems (OLTP)	*Decision Support Systems (Data warehouse-OLAP)*
Purpose	Automated day-to-day operation.	Information retrieval and analysis of detail historical data.
Structure	RDMS – Relational Database Management Systems	RDMS – Relational Database Management Systems
Data Model	Designed to work with relatively small pieces of information, therefore, normalized	Works with huge bocks of information, therefore, multi- dimensional. Mostly denormalized.
Access	Information System supported queries (SQL)	SQL plus business analysis extensions
Types of data	Data that runs the business	Information to analyze the business
Condition	Frequent updates in real time. Data incomplete and constantly changing. Often contain duplicate entries or reverse transactions Users can enter data. Historical data are missing	Historical, descriptive, summarized. Users never allowed to enter or update data.
Data Volumes	Gigabytes	Gigabytes,/terabytes
Implementation Cost	Moderate to expensive	Moderate to extremely expensive
Others	Usage patterns are relatively predictable.	Usage patterns not quite stable

Traditional OLTP systems, designed to handle day-to-day mission critical and access through simple query operations. They are very good at putting data into the databases quickly, safely and efficiently but not very good at delivering meaningful analysis. Retrieval of facts for a typical ad hoc report takes too long to specify and convey to an IMS (information management systems) organization with even longer wait for a result. Operational systems therefore can not serve as repositories for facts and historical data for business analysis. In the world of transaction processing systems, more demands are made of information system than questions asked of the data. The occasional queries are typically limited to locating a particular record in the information system and preparing it for updates, or performing simple aggregations.

Decision-makers on the other hand generate a breed of questions unlike those geared to transaction processing. Such queries originate from the business need to analyze and process data in order to draw conclusions. These questions are usually complex and typically cover dimensions not relevant to OLTP systems, such as time period, product families or region of the world (Adriaans & Zantinge 1996). As data warehouse is only applicable to decision support system environment, it is of little use to a firm that has a lack of integration in the operational environment. Nearly all organizations have built a collection of legacy applications over a period of many years. These systems are disorganized and typically old mainframe applications designed to do a specific job. Although they provide the backbone for operational processing, they are unintegrated because they were previously built separately from each other. An attempt can be made to integrate operational applications by producing data and process models that show the complete information flow requirements of the organization.

4 TYPES OF DATA IN THE DATA WAREHOUSE

We can think of the data warehouse as a collection of historical transactions and summarization. It may also be thought of as a giant spreadsheet that sits on a big computer. The data warehouse holds different flavors of data. The following represents a common sampling of the type of data contained in the data warehouse (Corey *et al.* 1998).

- Transaction downloads from operational systems that are time-stamped to form historical record.
- Dimensional support data (client, suppliers, materials, equipment, human resources time, cost, techniques that worked etc.).
- Table to supports the joining of dimensional data and numeric facts relating to this data.
- Summarization of transactions (e.g., daily production by departments). These are in actual facts preemptive queries; the data is aggregated when it is added to the data warehouse rather than when a user requests for it.
- Miscellaneous coding data
- Metadata which represents data about the data. This category might include sources of the warehouse data, replication rules, rollup categories and rules, availability of summarization, security and controls, purge criteria, logical and physical data mapping.
- Event data sourced from outside services, such as demographic information collated into the geographic areas in which a company operates.

5 OPERATIONAL DATABASE VS. DATA WAREHOUSE DESIGN

Operational system design and creating databases to serve operational purposes differ significantly from the goals of data warehouse design as we discussed in earlier sections. When creating a database for an online operational system in a relational model, the concern is with quick response time and efficient data storage and therefore the data model created must be in the third or higher normal form. As data is moved from operational systems into the warehouse, it goes through a process known as systematic denormalization that violates all the rules relational database architectures apply when modeling most systems. Systematic denormalization is carried out to enhance the performance of the warehouse by reducing *join* (an SQL operator) operations that are resource intensive. Compared to single table statements, *join* operations have the following characteristics (Corey *et al.*1998):

- Consume significantly more CPU;
- Require more temporary work space (on disk and in memory) for sorting;
- Require temporary tables for holding of intermediate results;
- Perform I/O since at least one I/O is required per table in the query

Ironically, when implemented, data warehouse will assume expansive size, does not necessitate dynamic updates, and is used primarily for analytical rather than repetitive processes. Hence, normalization is not only futile but is in fact counterproductive towards factors like performance and ease of use when it comes to decision support applications. The numerous join processes generated between normally distributed tables causes slow and cryptic query systems (SQL based) with non-instinctive application logic for Data Warehouse users. The Nested Relational Model is presented as an ideal solution (Inmon 1992b).

6   DATA ARCHITECTURE PERSPECTIVE OF DATA WAREHOUSING

In this section, we will look at the components that constitute the data warehouse, a data architecture perspective and information flow process in a typical data-warehousing environment. The data warehouse database is a combination of many different components, including the following (Corey *et al.* 1998):

- Operational data source
- The staging area
- The data warehouse
- The subject data marts
- OLAP Server(s)

- Reporting tools
- Matadata repository
- Monitoring and administration of the warehouse.

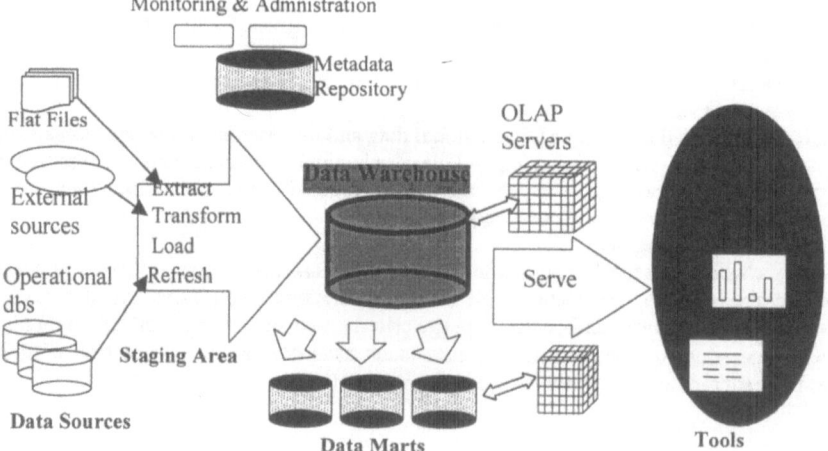

Figure 2. A Typical Data Warehousing Environment Architecture (adopted from Chaudhuri & Dayal 1996)

The architecture perspective of the data warehouse, includes tools for extracting data from multiple operational databases and external sources; for cleaning, transforming and integrating this data; for loading data into the data warehouse; and for periodically refreshing the warehouse to reflect updates at the sources and to purge data from the warehouse, perhaps onto slower archival storage. The data is initially extracted from the source such as operational data or flat files, through the staging area, and then loaded into a data warehouse using various third party loaders such as SQL* Loader and data warehouse loading tools. The warehouse is then used to populate the various process-oriented data marts and OLAP servers. The entire data warehouse then forms an integrated system that can serve the decision-maker reporting and analysis requirements of the user community (Chaudhuri & Dayal 1996).

The staging area is a set of database tables that will be used to receive data from the operational data source. This accelerates the manipulation and of the data in the relational database in the operational environment and loading into the data warehouse. The data structure in the data warehouse simplifies the enterprise's data while still retaining a non-process oriented database structure. In addition to the main warehouse, there may be several departmental data marts. Data in the warehouse and data marts is stored and managed by one or more warehouse servers (OLAP servers), which present multidimensional views of data to a variety of front end tools, namely, query tools, report writers, analysis tools, and data mining tools. Finally, there is a repository for storing and managing Metadata, and tools for monitoring and administering the warehousing system. (Inmon 1992a, Chaudhuri & Dayal 1997)

7 DATA WAREHOUSING AND CONSTRUCTION INFORMATION MANAGEMENT

Normally the operational database is the central storage area of all the data needed to serve each business function of the entire organization. Today, many construction organizations are moving toward the new technology offered by client server architecture to centralize their database at one location and decentralize their decision-making initiatives. In a typical construction management information system the database is essentially normalized to shrink the overall database size, easing updates, and establish program/data independence. As a result they are not able to support online analytic processing for strategic decision making.

Incorporation of historical, projected and derived data is an important process in the implementation of data warehousing solutions in any construction organization. Historical data is most often compiled from existing operational data, projected information usually comes form spread sheets or external data feeds and derived data is computed by OLAP (online analytic processing) server's calculation engine. By integrating all three types of data, construction workers will be able to make decision support operationally oriented process that looks forward as opposed to a system that reports only what has happened in the past. Support for historical data alone limits an operational system to only *what-happened* questions. Support for projected and derived data enables a system to additionally support what-if and what-next analyses. The ability of an OLAP server to deliver historical, projected and derived data side by side enables forward looking applications that provide significant value to end users that are otherwise impossible to deploy.

In a typical construction organization, the data will be gathered from vast amount of valuable information contained in operational sources such as accounting payroll, cost estimates, company and project finance, material inventory, equipment inventory, human resource data, and contract data. These data will then be transformed into a single integrated, subject-oriented database. Project engineers, architects, suppliers, construction managers and all personnel involved in a construction venture, can then query and analyze the information in the warehouse to derive informed business decisions in ways that were not possible before. A typical construction data warehouse will contain a large volume of historical data that will support more strategic decision support needs such as long-term trend analysis in project cost, project duration, tender quotations, experience derived from past projects etc. Building managers, for example, can compare alternative sites, building types, design and construction schedules and possible types of tenants or buyers in a local real estate market and be able to project expenditure and income streams to enable them to develop a more precise discount cash flow.

8 OVERALL BENEFITS OF IMPLEMENTING DATA WAREHOUSING IN A CONSTRUCTION ORGANIZATION

The overall benefit of data warehousing in any construction organization, view of the presentation in this paper, will be to serve as a distinct centralized repository for online transaction processing systems in the organization. This data may contain extracts of vital business data from a variety of corporate databases, which can be analyzed and used as a strategic competitive weapon. Successful data warehouse implementation in a construction organization will increase the productivity of construction managers, project developers, and the organization as whole. The inherent flexibility of online analytic processing (OLAP) systems means business users of OLAP applications can become more self-sufficient (Roussopoulos 1995). Managers are no longer going to depend on information technologists to make schema changes to create joins. Perhaps more importantly, data warehouse capabilities will enable managers to model problems that would be impossible using less flexible systems with lengthy and inconsistent response times. More control and timely access to strategic information equal more effective decision-making.

Considering the fact that the construction industry is rapidly becoming a global industry, it is expected that many construction organizations are going to commit their effort to achieve tremendous benefits by implementing data warehousing technology in their organizations. The reasons for this are not far fetched from the numerous advantages inherent in data warehousing technology discussed in this paper: These interesting benefits can more than can compensate for the investment made in the implementation of a workable data warehouse.

By giving construction employees robust access to information about customers, markets, suppliers, financial results etc., they will be enabled to strategically learn from the past, adapt in the present, and position for the future. A successful implementation of data warehousing will increase the productivity of construction managers, and their organization as a whole.

Online analytic processing (OLAP) capabilities of data warehousing will enable construction managers to model problems that would otherwise be impossible using less flexible systems with lengthy and inconsistent response time. Thus, construction organizations will be able to compete across time and learn from the past, adjust to the present, and position for the future. For optimally utilizing the concept of data warehousing, however, fundamental differences between Database and Data Warehouse such as current vs. historical data, large volume vs. very large volume data, mission critical vs. decision support application, etc. must be understood. Rather than adopting RDBMS for transaction based as well as information based applications, a clear distinction between DBMS (Database Management System) for the former and DWMS (Data Warehouse Management System) for the latter is recommended (The Data Warehouse 1998).

9 ORGANIZATIONAL ISSUES

Data warehouse developmental efforts, including attempts for improvements and changes, mandates considerations of how data warehousing will be supported in the construction organization. There will be the need for specific decision regarding who will maintain the database, networks, third party OLAP tools and programming and any other decentralized applications. Because the effectiveness of data warehouse depends on the willingness of users to share data, this coordination can be difficult. Traditionally, this type of responsibility rests with the data processing or information system (IS) entity within the organization. For success of data warehousing in a construction organization the administration function must be separated into a new unit that has authority across the entire organization. In any of this organizational structure, the enterprise nature of data warehousing requires that stakeholders be integrated into the functional system. A typical approach currently used in most organizations practicing data warehousing is to create a data warehouse coordinating group represented by all the various units of the organization (Kimball & Strehlo 1995).

Despite the convincing potential of data warehousing in construction information management, the decision to invest in this direction must be done with care since data warehousing is not always the most cost-effective or even the best solution for an organization. It may sometimes be advisable to start by building summarized reporting structure in the OLTP database. These structures can eventually be ported to the warehouse. Prior to any system development effort, it is also important to undertake careful financial analysis in terms of cost and benefits to determine the viability of the proposed investment in data warehousing. The costs are straightforward to estimate in a data warehousing environment as in any online transaction processing environment, however benefits are far less tangible because in most cases the exact target audience of the system is rarely known. In general, the decision to implement a data warehouse should be based on the following questions (Corey *et al.* 1998):

❑ Does it give us competitive advantage?

❑ Does it improve the bottom line?

❑ Will it deliver on all its promises?

❑ Will it deliver on time?

❑ What will be the risk if it is not implemented?

❑ What will be the risk if it is implemented?

❑ Will it deliver on budget?

It must be emphasized that while it is important to discuss the benefits of a planned warehouse, it is usually impossible to quantify these benefits in dollar terms. In many organizations that are currently practicing data warehousing technology, the decision to construct the warehouse is frequently an act of faith.

10 CONCLUSION

Most construction organizations have accumulated vast amount of database over the years through the normal daily transaction processing activities. These databases have, in most cases, not been designed to store historical data or respond to queries except to support all the applications used in the organization. Data warehouse represents another type of database, solely designed to support strategic decision that can be set up in any construction organization. Data warehousing technology can enable construction companies to consolidate information from diverse operational systems into one source for consistent and reliable information. This will give construction project managers the opportunity to bring wisdom and insight into their decision making process.

The premise of data warehousing, as presented in this paper, is to make large amount of information available to large community of end users in an organization. Presumably, the more users that are able to access data warehousing applications, the more value will be provided to the organization by the data warehouse. This implies that to maximize success, the online analytic processing (OLAP) server in the data warehouse must be made accessible from a wide variety of end-user tools (Arbor software 1997). As a data warehouse stores sensitive business information in a centralized location, the importance of data security and user management cannot be overemphasized.

While many commercial products and services exist, there are still several interesting avenues for research in order to adopt the most appropriate technique and customization to suite the needs of individual construction organizations. By providing the ability to model real business problems and more efficient use of people resources, data warehousing can provide the means to construction organizations to respond more quickly to market demands. Market responsiveness in turn often will yield improved revenue and profitability. A good data warehouse should provide the right data to the right people at the right time (DISC White Paper 1997) – which is the main purpose of any organizational information system.

REFERENCES

Adriaans, P. &. Zantinge, D. 1996. *Data Mining*. Reading, Massachusetts: Addison: Wesley Longman Limited,

Chaudhuri S. & Dayal, U. March 1997. *An Overview of Data Warehousing and OLAP Technology* Technical Report MSR-TR-97-14 Microsoft Research, Advanced Technology Division Microsoft Corporation: Redmond, WA.

Corey M., Abbey M.& Abramson I. 1998. *Oracle 8 Data Warehousing-A practical Guide to Successful Data Warehouse Analysis*. ORACLE Press Edition. Berkeley, California: Osborne/MacGraw-Hill.

Dynamic Information Systems Corporation (DISC) White Paper 1997. *Bringing Performance to Your Data Warehouse:* http://sun2.dic.com/dwhper.html.

Inmon W.H. 1992a. *Building the Data Warehouse*. New York, NY: John Wiley:

Inmon W.H. 1992b. *Rdb/VMS: Developing the Data Warehouse*. New York, NY: John Wiley.

Inmon W.H. 1992c. *Using the Data Warehouse*. New York, NY: John Wiley.

Kimball R. & Strehlo 1995. Why decision support fails and how to fix it, reprinted in *Sigmod Record*, 24(3).

Kimball, R. 1996. *The Data Warehouse Toolkit*. New York, NY: John Wiley. NY.

Roussopoulos, N. et al. June, 1995. The Maryland ADMS Project: Views R Us. *Data Eng. Bulletin*, Vol. 18, No.2.

Rezgui, Y., G. Cooper & P. Bradon, July 1998. Information Management In a Collaborative Multiactor Environment: The COMMIT Approach. *Journal of Computing in Civil Engineering: 136-142.*

The Data Warehouse: Achieving Better Decisions Faster, 1998. *Database and Network Journal*: An International Journal of Databases and Network Practice Vol. 28,.No3: 3-6.

The Role of the OLAP Server in Data Warehousing Solution. *Arbor Software White Paper*, 1997 Sunnyvale, California: http://www.aborsoft.com.

Veshosky, D. January/February 1998. Managing Innovation Information in Engineering and Construction Firms. *Journal of management in engineering*:58-66.

Zhuge, Y., H. Garcia-Molina, J. Hammer & J. Widom, 1995. View Maintenance in a Warehousing Environment, *Proceeding. of Sigmod Conference.*

Artificial Intelligence and Mathematical Methods in Pavement and Geomechanical Systems, Attoh-Okine (ed.)
© *1998 Balkema, Rotterdam, ISBN 90 5809 028 0*

Probabilistic graphical networks in pavement management decision making

N.O.Attoh-Okine & Alexander Kwasi Appea
Department of Civil Engineering, Florida International University, Fla., USA

ABSTRACT: Decision making under uncertainty is common in pavement management decision making problems. In finding solutions to these problems, the main objective is to select an appropriate (or optimal) decision alternative in the face of uncertain environment. Good decision support information improves pavement decision-maker's insight on the potential for various decision alternatives to achieve desired outcomes. In this paper, probabilistic graphical networks (Bayesian Networks, Influence Diagrams and Valuation-Based Networks) are introduced as a useful class of methods for structuring and solving pavement decision making under uncertainty.

1 INTRODUCTION

A pavement management system (PMS) is designed to provide information and useful data analysis so that highway managers and engineers can make more consistent, cost effective and defensible decisions related to the preservation of a pavement network (AASHTO, 1996). PMS, in its broadest sense, encompasses all the activities involved in the planning, design, construction, maintenance, and rehabilitation of a public works program.

In a PMS, the problem setting often involves a large number of uncertain, interrelated quantities attributes and alternatives based on information of highly varied quality. Pavement management is generally described and developed at two levels: the network and project level. The primary differences between the network and project level decision making tools include the degree or extent to which the decision is being made and the type and amount of data required. Generally, the network level decisions are concerned with programmatic and policy issues for an entire network whereas the project level decisions address the engineering and economic aspects of pavement management.

The decision aid technique is also known as intelligent decision support, which is based on human knowledge understood as a family of classification patterns related to a specific part of a real or abstract world. When the knowledge is gained in the process of learning by experience, it is induced from empirical data. The data are often presented as a record of objects described by a set of multivalued attributes like features, variables, characteristics, conditions, and so on. The objects are associated with some decisions (actions, opinions, classes, diagnoses, etc.) taken by an expert. Such a record is called an information system.

The traditional decision support techniques, which have long been used in PMS, include a decision tree called linear programming methods. A decision tree is a visual display of the structure of a decision problem. It looks like a tree with branches spreading out from nodes. Linear programming is a mathematical technique used to optimize resource allocation when confronted with certain side constraints that limit the range of choices. Although decision tree and linear programming methods can solve some problems in the pavement management system, many limits exist for their applications. The linear programming method can only be applied to the problems with linear properties. The decision tree and linear programming methods also rely on some other techniques to address uncertainties in decision-making problems.

A probabilistic graphical model is a graph where the nodes represent variables and the arcs (directed or undirected) show dependencies between variables. These are then used to define/obtain a mathematical form for the joint or conditional probability distribution between variables. Probabilistic graphical forms come in various forms: Bayesian networks used to represent casual and probabilistic processes, data flow diagrams used to represent deterministic computations, influence diagrams used to represent decision processes, undirected Markov networks for hidden causes, and valuation and networks for deterministic computation and assymetric decision making.

This paper presents an overview of selected probabilistic graphical models and their potential application to pavement decision making.

2   BELIEF NETWORK

Belief Networks are an expressive graphical language for representing uncertain knowledge about the casual and associational relationships among variables in complex systems. They are directed acyclic graphs (without directed cycle) in random variables and arcs that represent dependencies. The graph serves as both a visual representation of the model and guide to efficient probability computation algorithms. The graphs serve as both a visual representation of the model and guide to efficient probability computation algorithms. The belief networks present systems in which uncertain information is available on a set of mutually dependent objects by a prior probability distribution to each object and with an expression of the strength and character of dependency between one pair of objects. Using the information, posterior probability distributions are calculated for each object (Jensen, 1996). The networks can also be used as inference mechanisms to answer probability queries though the representation encodes the joint probability distribution for its domain, yet includes a qualitative structure that facilitates communication between a pavement engineer manager and a platform-incorporating probabilistic model.

A belief network consists of qualitative and quantitative representations. The qualitative representation is the acyclic graph $G = [V(G), A(G)]$, where $V(G) = (V_1 \dots V_n)$, $n \geq 1$ is a finite set of vertices and $A(G)$ is a finite set of arcs $\{V_i, V_j\}$, $V_i, V_j \in V(G)$. Each V_i in $V(G)$ represents random variables. The arc $\{V_1, V_J\} \in A(G)$ represents a direct influence or casual relationship of the variable. Nodes that are linked, directly or indirectly, are considered to have some relevance to each other (V_i and V_j). Let us consider a finite set of pavement condition variables $U = \{U_1, \dots U_n\}$, each with a finite domain and with a probability distribution $p(U)$. A directed acyclic graph D of a joint probability distribution $p(U)$ is a Bayesian Network of p of D constructed from p by the following steps (Geiger & Hackermann, 1998).

Assign an arbitrary construction order for the variables U_i, $U_2, \dots U_3$ to the variables in U. For each variable U_i in U identify a set
$C_I \subseteq \{U_1, \dots, U_{i-1}\}$ such that
$\{ U_i \} \perp \{ U_1, \dots U_{i-1}\} C_i \setminus C_i$
By the chaining rule it follows that
$P(U_1, \dots, U_n) = \Pi P(U_i \mid U_1, \dots, U_{i-1})$
$P(U_1, \dots, U_n) = \Pi P(U_i \mid C_i)$.
The joint distribution is represented by the network and can be used for computing the posterior of every pavement condition variable given a value to some other relevant variables in the network. The number of parameters that a network requires and the complexity depend on the

construction order. Construction orders include: causal-and-effect, and time-order relationships relevant to the impact on traffic on the pavement.

Figure 1a shows no independence. Directly connected nodes always relate to each other.

Let Traffic be represented by T and Pavement Condition PC.
T ∈ { low traffic ' (It) ', medium traffic ' (mt) ', heavy traffic ' (ht) '}
PC ∈ { Below acceptable condition ' (bac) ', Normal Acceptable Condition
' (nac) ', Above Acceptable Condition ' (aac) '}.

Table 1. Probability of Traffic Condition

$P(T=' It ') = 0.80$
$P(T=' mt ') = 0.15$
$P(T=' ht ') = 0.05$

Using Figure 1a, the following expression can be developed.

Table 2

⇓PC T⇒	' It '	' mt '	' ht '
P(PC= bac)	0.96	0.88	0.60
P(PC=nac)	0.03	0.08	0.25
P(PC=aac)	0.01	0.04	0.15

Using the Product Rule

Table 3. $P(PC, T) = P(PC \mid T) \cdot P(T)$

⇓PC T⇒	' It '	' mt '	' ht '
' bac '	0.768	0.024	0.008
' hac '	0.132	0.012	0.006
' aac '	0.035	0.01	0.005

Table 4. Marginalization can be computed.

⇓PC T ⇒	' lt '	' mt '	' ht '	Total
' bac '	0.768	0.024	0.008	0.80
' hac '	0.132	0.012	0.006	0.15
' aac '	0.035	0.01	0.005	0.05
TOTAL	0.935	0.046	0.019	0.05

$P(PC) = 0.935 + 0.046 + 0.019$

$P(T) = 0.8 + 0.15 + 0.05$

The Bayes rule

$$P(T/PC) = \frac{P(PC/T)\ PCT}{P(PC)} = \frac{P(PC,T)}{P(PC)}$$

Table 5

⇓ PC T ⇒	' lt '	' mt '	' ht '
' bac '	0.768/0.935=0.821	0.024/0.046=0.522	0.008/0.019=0.421
' nac'	0.132/0.935=0.141	0.012/0.046=0.261	0.006/0.019=0.316
' aac'	0.03/0.935 = 0.037	0.015/0.046 = 0.217	0.005/0.019=0.263

Figure 1b shows marginal independence. The nodes do not have a common ancestor (cause). They are said to be marginally independent, if the status of their common descendents (effects) is unknown. Belief in data collection is irrelevant to previous conditions if pavement performance is unknown. Figure 1b sometimes refers to a converging connection.

Let Data Collection = (DC), Previous Condition = (PreC), and Pavement Performance = (PR).

The conditional independence shows that

$P(DC \mid PreC, PP) = P(DC \mid PP)$

Figure 1c shows conditional independence. Nodes that share a common ancestor or have intervening nodes along the direction of the arrow (cause-effect-effect). When the status of their mutual ancestor or intervening node is known, belief in one node will not affect the belief node.

Figure 1d shows complete independence. Nodes that have no connection are entirely irrelevant to each other.

Independence

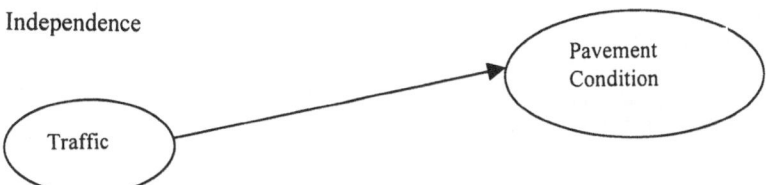

Figure 1a.　　Belief of Traffic Impact on Pavement Condition.

Conditional
Independence

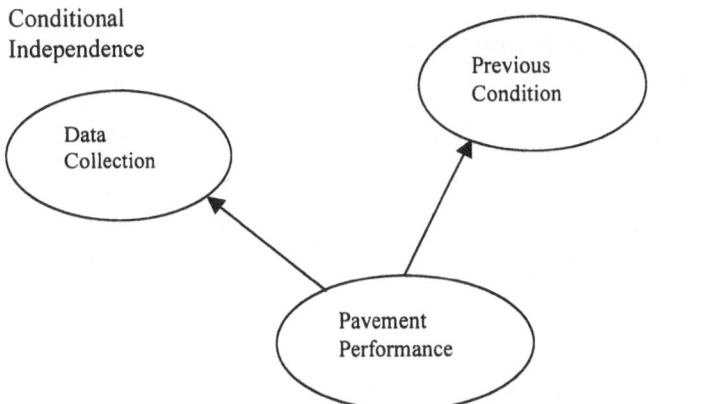

Figure 1b.　　Belief in Data Collection is irrelevant to Previous conditions if the Pavement Performance is unknown.

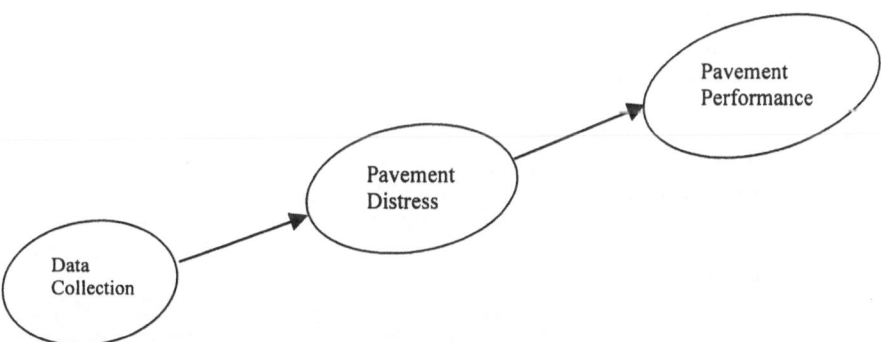

Figure 1c.　　Belief in Pavement Performance is irrelevant to the Data Collection procedure of a database if Pavement Distress is known.

Figure 1d.　　Belief in Data Collection is never relevant to the belief in Traffic Impact.

$P(DC,T) = P(DC) P(T).$
$P(DC | T) = P (DC) \quad DC \perp T$
$P(T | DC) = P(T), \quad T \perp DC$
$P(DC, T) = P(T | DC) P(DC) = P(DC) P(T)$
$P(DC, T) = P(DC | T) P(T) = P(T) P(DC)$

Node V_i contains:

- A vector of outcomes that can be defined as inputs, or they may depend on outcome values of other nodes.
- A prior probability distribution with outcomes given.
- A sign indicating direction of changes.
- A posterior probability distribution.

A link transfers information between two nodes. Links define a link matrix and an outcome link which present a functional relationship. An example of belief networks is presented in Figure 2. Figure 2 shows a subnetwork of PMS decision-making variables. In Figure 2, it is assumed that Payoff (P) depends on Budget (BL) and pavement performance. Pavement performance (PP) depends on data collection (DC) at previous conditions of the pavement (PC). The data collection (DC) depends on the validity of the data (VD) and reliability of equipment (RE). The joint probability is given by the following formula $Pr(V_i/S_i)$, where S_i is the parent of node V_i (variables)- (Binary variables).

$Pr (VD, RE, DC, PC, PP, BL, P) = Pr (VD) * Pr (RE) * Pr (PC) Pr(DC | VD, RE) * Pr(C) * Pr (PP | DC, PC) * Pr(BL) * Pr(P | BL,PP)$ \hfill (1)

In Figure 2, the Belief Network implies a factorization of the joint distribution of the seven variables which embody a number of conditional independence.

$P \perp (VD, RE, DC, PC) | BL, PP$

$PP \perp (VD, RE) | DC, P$

$\perp$ - denotes conditional independence.

Equation 1 is known as a chain rule. Note for variables with n incoming arcs, 2^n probabilities have to be assigned, and for variable with zero predecessors, only one probability has to be specified, namely the prior probability . If there are n binary random variables, the complete distribution is specified by 2^n-1 joint probabilities. Thus, Figure 2 has 128 values. Fortunately, Belief networks have a built-in independence assumption that helps to reduce these specifications drastically.

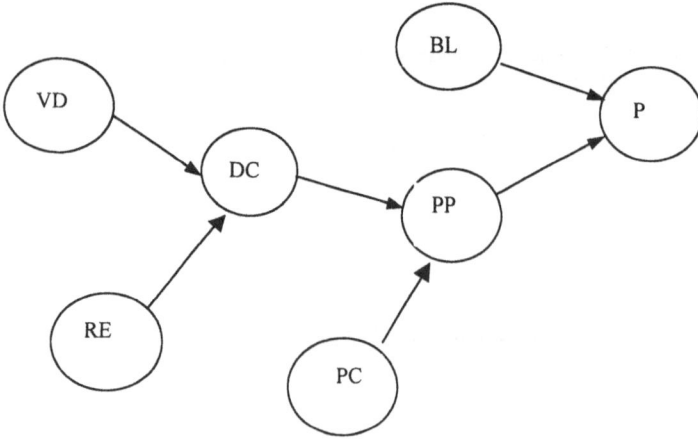

Figure 2.   Prototype Belief Network.

3   INFLUENCE DIAGRAMS

Influence Diagrams are graphical structures for representing decision problems. An Influence Diagram represents all the components of a decision problem, i.e., the decision objectives (modeled using utility functions), decision problem, values, uncertainties, as well as the relationships among them (Figure 3).

The detailed data about the variables are stored within the nodes, so that the diagram is compact and focuses attention on the relationships among the variables. Moreover, Influence Diagrams enable symbolic modeling of decision problems without necessarily requiring in-depth knowledge of their underlying theory, hence their potential utility and appeal to analysts, experts and decision-makers. And last but not least, the solution algorithm for Influence Diagrams is more direct and can lead to considerable computational savings (Shachter, 1986), especially where the problem structure is symmetric. The Influence Diagram model has a convenient structure for computer manipulation and solution procedures. As Kirkwood (1992) points out, algebraic formulation methods allow Influence Diagrams to represent decision models that are too complex to handle with Decision Trees. However, Influence Diagrams are intrinsically symmetric and do not graphically depict asymmetric problems. The reduction algorithm has been widely tested and is currently the acceptable method for the solution of Influence Diagrams (Shachter, 1986; Shachter, 1988; Clemen, 1991; Attoh-Okine & Roddis, 1998; Amekudzi & Attoh-Okine 1997). This algorithm takes advantage of the inherent symmetry of Influence Diagrams. Specifically, the reduction algorithm may not be advantageous for problems in which most of the computational savings are gained through asymmetric processing. Comprehensively, therefore, Influence Diagrams may be a more practical tool for decision modeling under uncertainty. They have been found to improve communication among decision-makers and analysts about key dependencies in a decision under uncertainty. In summary then, Influence Diagrams are intuitive; they are better able to model larger decision problems, they emphasize the critical relationships among various variables, and communicate a complete value and probabilistic description of the problem from the decision-maker's viewpoint (Shachter, 1986; Shachter, 1988; DPL, 1995).

3.1   *Topology of Influence Diagrams*

Topologically, an Influence Diagram is a finite non-cyclic graph made up of directed arcs (arrows) linking four kinds of nodes: decision nodes, deterministic nodes, chance nodes, and value nodes.

3.2   *Nodes*

Nodes represent variables. A node represents a choice among a set of alternatives. Each node contains a list of the possible values of the variables that the node represents. Chance or random variables are depicted by circles; decision variables by rectangles, deterministic nodes by concentric circles and value or utility nodes by rounded rectangles. Each chance node contains a probability distribution for its variable X for each configuration of its predecessor nodes. The probability distributions may be obtained from subjective assessments by experts, from maintenance records, statistical databases, or experimental data. Each decision node contains a number of decision options and represents the choices available to the decision-maker. Deterministic nodes may be thought of as a special kind of chance node in which all the probabilities happen to be zero or one: i.e., a deterministic node has a number of states and at any point in time, there is only one state (with an associated probability of 1) that may be assumed by the deterministic node. The value node may be viewed as a special kind of chance node whose value is needed to answer the question of interest to the analyst. This value node contains a mapping that specifies the value of its variable X given values of all its predecessor nodes (Shachter, 1986; Agogino et al., 1992).

3.3   *Arcs*

Arcs linking two nodes indicate some kind of influence of one node on the other. There are two kinds of arcs: informational and conditional arcs. Known as conditional arcs, arcs into chance or value nodes indicate that there may be probabilistic dependence. Arcs into decision nodes are called informational arcs and simply imply time precedence; they indicate that information from the predecessor nodes must be available at the time of decision (Shachter, 1986: Shafer, 1992) Figure 3 illustrates fundamental node relationships in an Influence Diagram.

3.4   *Evaluation of Influence Diagrams*

In order to evaluate an Influence Diagram, there must be a question to be answered, i.e., some random variable(s) whose distribution(s) must be determined. The associated chance node is singled out as the value node. This value node then represents the objective to be optimized (maximized or minimized) in expectation. There may be single or multiple variables associated with the value node. The variable(s) associated with nodes having arcs into the value node are the attributes of the decision-maker's utility function. The random variable of the value node needs to be calculated in expectation. This expected value represents the utility of the outcome to the decision-maker. If there are decisions to be made then the expected utility may be used to compare alternatives. Given the state of information at the time of the decision, the alternative(s) selected should maximize the expected utility of the resulting outcome (Shachter, 1986).

Although a variety of algebraic algorithms are available for evaluating Influence Diagrams, the reduction algorithm had been widely tested and is currently used in the solution of Influence Diagrams (Shachter, 1986; Clemen, 1991). The reduction algorithm has been used in developing several types of commercially available Influence Diagram software packages such as DPL^{TM} (Decision Programming Language), $InDia^{TM}$ (Influence DIAgram processor) and $Demos^{TM}$. These software tools generally run on widely used IBM compatibles and Macintoshes with graphical user-interfaces, and are reasonably easy to use once their basic decision analytical concepts are understood (Maxwell, 1996).

3.5   *Example*

The relationship between variables, decisions, and outcomes during a maintenance decision and rehabilitation at the project level is difficult to understand even for most experienced pavement engineers. In pavement maintenance decision-making, alternatives are not generated and

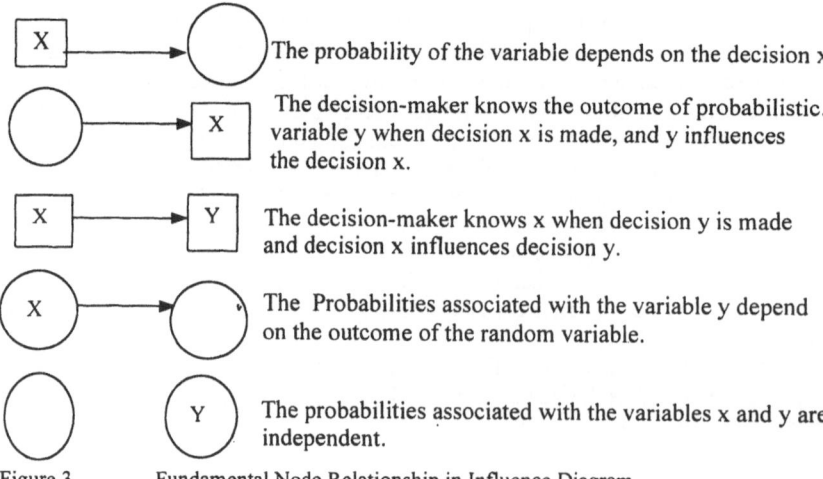

Figure 3.   Fundamental Node Relationship in Influence Diagram.

126

logically composed. Pavement engineers and policy makers assess the state of the pavement, resources and budget level and make sets of decisions that match the maintenance problem they perceive. Given the importance of uncertainties and utilities of the pavement (user costs/agency cost) in maintenance decisions influence diagram seemed most appropriate for representing and solving these types of problems. The advantages to this approach are that pavement utilities, incorporating preference and risk attitudes, are explicitly modeled. The influence diagram representation is useful in providing the decision-makers (pavement engineer and upper management) with the clear view of the model variables and the relation between them.

Figure 4 show the Bayesian influence diagram of a pavement maintenance model. The influence diagram is based on the author's experience and approaches used in pavement maintenance manuals. Ideally, the data for the model will come from PMS database and the probability for the variables can come from variety of sources, the distribution can be derived from the database, defined subjectively by the decision maker. The diagram contains ten nodes. Two decision nodes, seven chance nodes and one payoff node. Pavement condition variables (cracks, rutting and deflection) and the maintenance alternatives are one inch overlays, three inch overlays and routine maintenance. Apart from the utility node, all the chance nodes were assigned equal chance (this for illustration purposes), the probabilities can be updated as more data become available. Environmental outcomes are good and severe. Pavement conditions are good and fair, impact on traffic are fair and bad, cost items are classified as high or low, rates of initiation are also slow and fast and percentage of section below the threshold are assigned high or low.

The model can be evaluated by node reduction algorithm (Shachter, 1986), and the objective is to find the optimal policy.

4   VALUATION-BASED SYSTEMS FOR PAVEMENT MANAGEMENT DECISION MAKING

The semantics of VBS can represent and solve Bayesian decision and optimization problems (Shenoy, 1991; Attoh-Okine, 1995; Attoh-Okine, 1997). VBS networks are generally designed

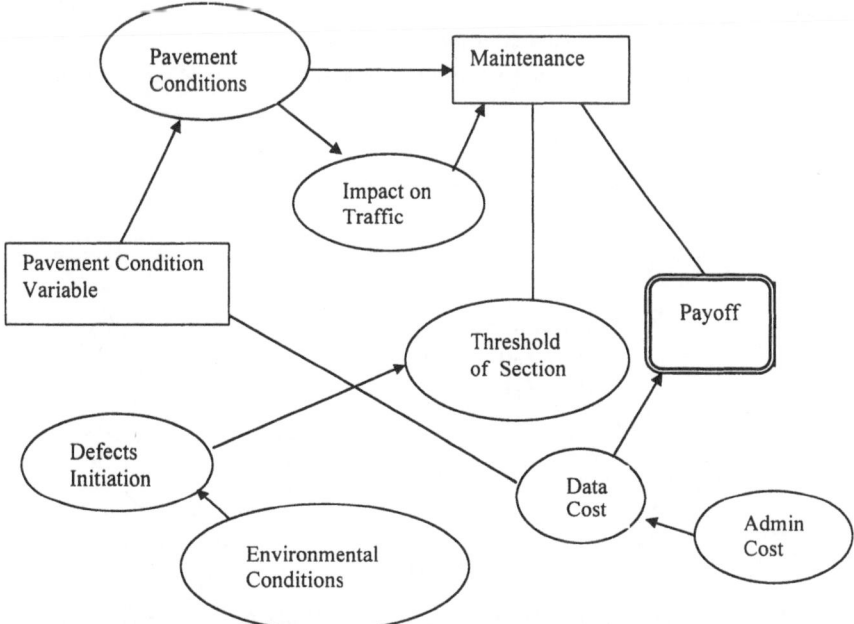

Figure 4. Bayesian influence diagram of a Pavement Maintenance Decisions.

for both asymmetric and symmetric problems, types which are encountered in infrastructure decision analysis.

A valuation-based system representation of a decision problem consists of a finite set of decision variables X_D, a finite set of random variables X_R, a finite set of random variables $\mathbf{X_R}$, a finite frame $\mathbf{W_x}$ for each variable X in $X_D \cup X_R$ a finite collection of payoff valuations $\{\pi_1 \ldots \pi_m\}$, a finite collection of potentials $\{\rho_1 \ldots \rho_n\}$, and a precedence relation $\rightarrow$ on X_D and X_R. Thus, a valuation-based system (VBS) for a decision problem can be denoted formally by the 6-tuple:

$$\Delta = \{X_D, X_R, \{W\chi\}_{\chi \in X}, \{\pi_1 \ldots \pi_m\}, \{\rho_1 \ldots \rho_n\}, \rightarrow\}.$$

To illustrate the difference between VBS and a decision tree, a canonical decision problem is presented and discussed. A canonical decision problem Δ_c consists of a single random variable D with a finite frame W_D, a single random variable R with a finite W_R, a single payoff valuation π for $\{D,R\}$, a single conditional potential ρ for R given $\{D\}$, and a precedence relationship $\rightarrow$ defined by $D \rightarrow R$.

The meaning of the canonical problem is as follows: the elements of W_D are, and the elements of W_R are states of nature. The conditional potential ρ is a family of probability distributions of R, one for each act $\mathbf{d} \in W_D$, that is, the probability distribution of a random variable $\mathbf{R=r}$ given that $\mathbf{D=d}$. The VBS representation or valuation network of a canonical decision problem is illustrated in Figure 5. Figure 6 shows the corresponding decision tree representation of the canonical decision problem.

In valuation networks, circular nodes represent random variables, rectangular nodes represent decision variables, triangular nodes represent potentials, and diamond-shaped nodes represent utility functions. The undirected edges linking variables to potential and utility functions denote the domains of these functions. The directed arcs between variables define the information constraint. If there are no random variables in the problem, the network reduces to an optimization problem.

4.1   *Variables and Frames*

A decision node is represented as a variable. The set of all possible values of variable X, denoted by W_X, is called the frame for X. Given a non-empty subset of $\mathbf{h}$ of variables, $\mathbf{W_h}$ denotes the cartesian product of $\mathbf{W_x}$ for X in h, i.e. $W_h = \mathbf{x}\{W_x \mid X \in h\}$. If h is a subset of variables, potential (or a probability function) α for h is a function α:

$W_h \rightarrow [0,1]$. The values of potential α are probabilities, and $\mathbf{h}$ is called he domain of α. If $\mathbf{h}$

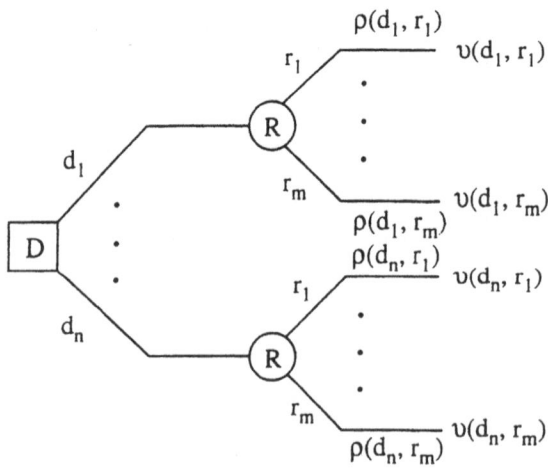

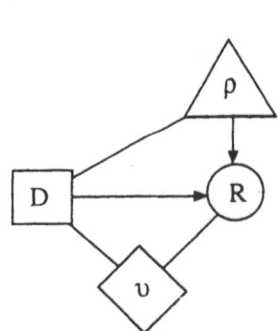

Figure 5.   Graphical representation of VBS.

Figure 6.   An equivalent decision tree representation of VBS.

is a subset of variables, a utility function υ for h is function: $W_h \to R$. The values of the utility function υ are utilities, and are called the domain of υ. If the set of variables **h** is empty, the following convention is adopted: for the empty set ϕ, there exists a single configuration, and we use the symbol ◆ to name that configuration $W_\phi = \{◆\}$.

For $r_i \in W_{\{Ri\}}$, r_i called a configuration, denotes a possible value of R_i and $W_{\{Ri\}}$, called the frame for $\{R\}$, r also denoted the configuration of R_i for $d_j \in W_{\{Dj\}}$, d_j called a configuration, denotes a possible action (decision) D_j, and $W_{(Dj)}, d_j$, called a configuration, denotes a possible (decision) Dj, and $W_{\{Dj\}}$, called the frame for $\{\mathbf{Dj}\}$, denotes the configuration of $\mathbf{D_j}$.

4.2 *Valuation*

Valuations are primitive in the VBS framework. Given a subset **h** of variables in **H**, there is a set $\mathbf{V_h}$ called valuations. Intuitively, a valuation for **h** represents some knowledge about the variable in **H**.

Let V denote the set of all valuations, i.e. $\mathbf{V} = \cup \{\mathbf{V_h} \,|\, \mathbf{h} \subseteq \mathbf{X}\}$. If σ is a valuation for h, then h is the domain of σ. The values of the utility valuations are utilities.

4.3 *Non-zero Valuation*

For each $h \subseteq X$, there is a subset $\mathbf{P_h}$ of $\mathbf{V_h}$ whose elements are non-zero valuations for **h**. Let P denote $\cup\{Ph \,|\, h \subseteq X\}$, the set of all non-zero valuations. Intuitively, a non-zero valuation represents knowledge that is internally consistent.

4.4 *Precedence Constraints*

Besides acts, events, probabilities and payoffs, an important ingredient of VBS representation are information constraints. Generally, four constraints are needed for the precedence relation:
(1) The transitive closure of $\to$, denoted by $>$, is partial order (irreflexive and transive) of $\mathbf{X} = \mathbf{X_D} \cup \mathbf{X_R}$. This is called the partial order condition;
(2) For any $D \in X_D$ and any $R \in X_R$, either $R > D$ or $D > R$. This is called a perfect recall condition;
(3) If there is conditional potential for R ($R \in X_R$) gives $h - \{R\}$, and there is a decision variable $D \in h$, then $D > R$; Suppose $h \subseteq X$, $R \in h$, $R \in X_R$ and ρ is a potential for h, ρ is called a conditional potential for R given $h - \{R\}$ if $\rho^{\downarrow h\{R\}}$ is vacuous potential.
(4) If there is potential for h and a decision variable $D \in h$, then $D > R$ for some random variables $R \in h$. The third and fourth conditions are called consistency conditions.

Given the meaning of the precedence relation $\to$, for any decision variable D and any random variable R, either R is known when decision D has to be made or not. This translates to either $R > D$ or $D > R$. Finally, the consistency conditions are dictated by the meaning of potentials. If the conditional probability distribution of random variable R depends on the act chosen by the decision maker at node D, then it must be the case that $D > R$. For example if the true configuration of R_i is known, before a configuration of D_j has to be chosen, then $R_i > D_j$; and if a configuration of D^1_j ha to chosen before a configuration of D^2_j, then $D^1_j > D^2_j$.
In consequence of the precedence relation, a specified set of precedence constraints may result in the non-existence of a solution of a decision problem. For instance:
If $\mathbf{D_j} > \mathbf{D^1_j}$, $\mathbf{D^1_j} > \mathbf{D^2_j}$ and $\mathbf{D^2_j} > \mathbf{D_j}$
This means that a condition has to be imposed on the precedence relation (Greve et al. 1992).

4.5 *Solution of VBS*

The solution method for VBS is based on the fusion algorithm (Shenoy, 1991). The basic idea is to successively delete all variables from VBS. The sequence in which the variables are deleted must respect the precedence constraint. The following theorem describes the fusion algorithm for making inferences in VBS.

Theorem (Shenoy, 1991c):

Suppose $\Delta = \{X_D, X_R, \{W_\chi\}_{\chi \in x}, \{\pi_1 \ldots \pi_m\}, \{\rho_1 \ldots \rho_m\}, \rightarrow\}$, is a well defined decision problem. Suppose $X_1, X_2 \ldots X_k$ is a sequence of variables in $X = X_D \cup X_R$ such that with respect to the partial order $>$, X_1 is a minimal element of X, X_2 is a minimal element of $X - \{X_1\}$, etc. Then $\{(\otimes\{\pi_1 \ldots \pi_m, \rho_1, \ldots \rho_n\})^{\downarrow \varnothing}\} = \text{Fus}_{xk}\{\text{Fus}_{x2}\{\pi_1 \ldots, \pi_m, \rho_1 \ldots, \rho_n\}\}\}$

The fusion algorithm contributes to the solution efficiency of the valuation networks and helps to achieve local computation, compared to global solution in other decision analysis methods like decision trees.

A strategy is a choice of an act or each decision variable D as a function of the configuration of random variables $R > D$. This table is regarded as a function, and it is called a solution for that decision variable. Suppose $D \subseteq h$, υ is a valuation for h, ψ_D: $W_{h-\{D\}} \rightarrow W_D$ is used to denote the solution for **D**.

In solving VBS, two operators, the combination and the marginalization are used. The combination is a mapping $\otimes$ V xV $\rightarrow$ V such that:

 (i) If ρ and σ are valuations for r and h, respectively, then $\rho \otimes \sigma$ is a valuation for $h \cup r$;

 (ii) If either ρ or σ is not a non-zero valuation, then $\rho \otimes \sigma$ is not a non-zero valuation; and

 (iii) If ρ and σ are both non-zero valuations, then $\rho \otimes \sigma$ may not be a non-zero valuation.

$\rho \otimes \sigma$ is called the combination of ρ and σ. The combination depends on the type of valuations being combined. The combination of two-payoff valuations is a payoff valuation; the combination of two potentials is a potential, and the combination of a payoff valuation and a potential is a payoff valuation. The combination of two payoffs consists of point-wise addition. The combination of two potentials consists of point-wise multiplication, and the combination of payoff valuation and a potential consists of multiplication.

Marginalization: Suppose h and g are subsets of variables, and suppose g is a subset of h. For each $h \subseteq X$, there is a mapping $\downarrow h : \cup \{\upsilon_g | h \subseteq g\} \rightarrow \upsilon_h$, called the marginalization to h, such that, if υ is a valuation for g and $h \subseteq g$, then $\upsilon^{\downarrow h}$ is a valuation on h. Marginalization depends on the type of valuation being marginalized. The definition of marginalization depends on the type of variables being eliminated. If the variables being eliminated are random, marginalization is achieved by maximization (or minimization depending on the nature of the values of the payoff valuations) over the frame of eliminated variables.

The maximum expected utilities are obtained by:

  $(\otimes\{\pi_1, \ldots \pi_m\})^{\downarrow \varnothing}(\blacklozenge)$.

The optimal strategy σ^* that provides the maximum expected value of Δ is determined as follows:

  $(\pi \otimes \rho)^{\downarrow D}(d) = (\otimes\{\pi_1, \ldots \pi_m, \rho_1, \ldots \rho_n\}) \downarrow \varnothing (\blacklozenge)$.

Where π, ρ, and D refer to the equivalent canonical decision problem Δ_C.

4.6 Example

Figure 7 uses the decision tree to illustrate the application of the Markov decision process in Pavement Management Systems. In this example it is assumed that the pavement serviceability index is 3, and the only feasible action is routine maintenance only. For simplicity, three discrete levels of PSI are assumed good (greater than 3), fair (2 to 3), and poor (less than 2). Conditional on each outcome, appropriate alternative actions are routine maintenance, one-inch overlay, and three inch overlay. Figure 7b is a simplified decision tree of Figure 7a. In Figure 8, a valuation-network is used to represent the same problem. The original decision tree does not provide the probabilities and the utilities. Since the decision node R has only one value, it can be omitted from the valuation network. Although it is not clear from Kulkarni's (1984) example, it is assumed that the conditional probability of C_2 depends on C_1. This is not clear from the decision tree. Secondly, the payoff also depends on C_1. The valuation network assumes that all three alternatives (routine maintenance, 1" overlay, and 3" overlay) are available regardless of the value of C_1.

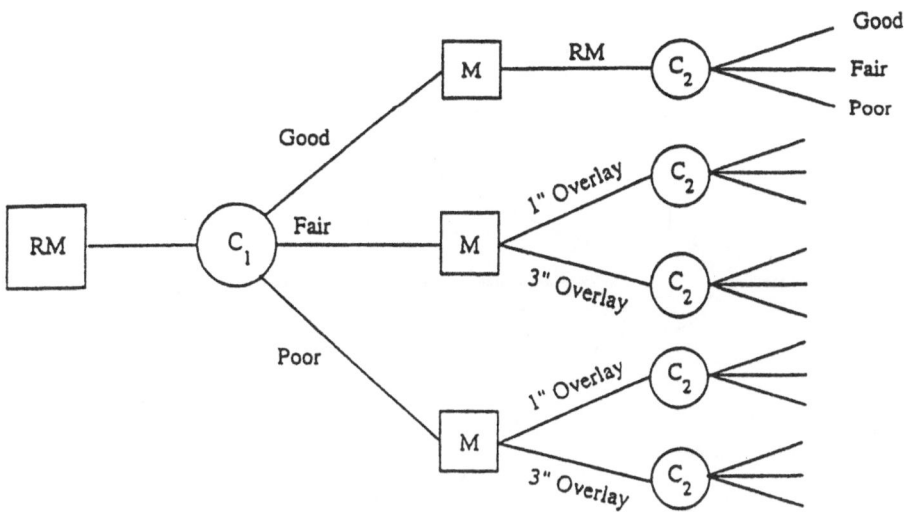

Figure 7a.   A decision tree in PMS (Kulkarni, 1984).

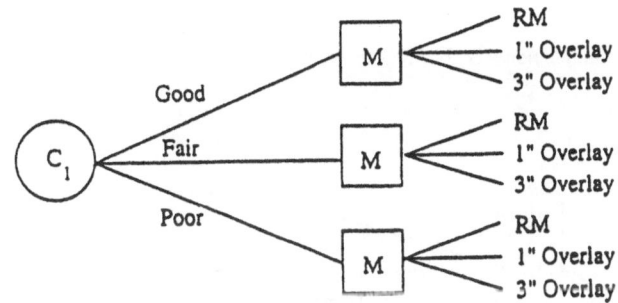

M   - Maintenance Decision
RM   - Routine Maintenance
C_1 - Conditional Probability Distribution (Year 1)
C_2 - Conditional Probability Distribution (Year 2)

Figure 7b.   A simply decision tree for Figure 7a.

The valuation-based system in Figure 8 is representative of the decision tree. The assumptions are important since numbers are omitted in the corresponding decision tree. The assumptions are important since numbers are omitted in the corresponding decision tree. The valuation network consists of two decision variables, RM (Routine Maintenance) with frame {RM} and M (Maintenance Activities) with frame {RM,1"overlay, 3" overlay}, and two random variables C_1(Pavement Condition in Year 1){good, fair, poor}. The precedence relation is R $\rightarrow C_1$, $C_1 \rightarrow$ M, M$\rightarrow C_2$. There are two potentials and one utility valuation in the network. Note that ν is a potential for C_1, and ρ is a potential for $\{C_1, C_2, M\}$. One can use the valuation network to compute the maximum expected utility value and an optimal strategy given by the solution of M.

The fusion algorithm Figure 9, avoids the computation on the frame of all the three variables. The maximum expected utility is $(\nu \nu \otimes ((\pi \otimes \rho)^{\downarrow \{C1,M\} \downarrow C1})^{\downarrow \varnothing}$. An optimum strategy is given by the solution for M with respect to $(\pi \otimes \rho)^{\downarrow \{C1,M\}}$

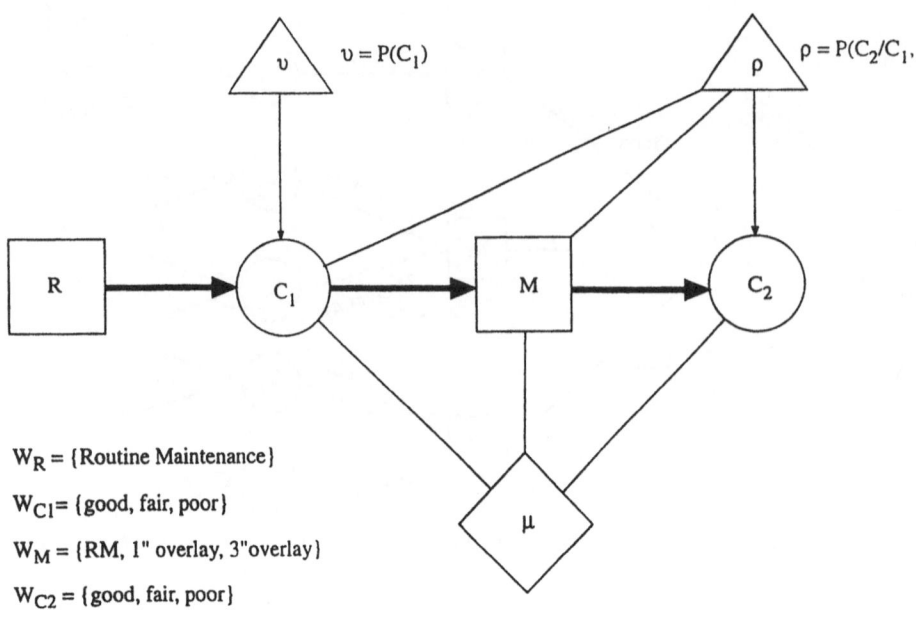

$v = P(C_1)$

$\rho = P(C_2/C_1,$

$W_R = \{\text{Routine Maintenance}\}$

$W_{C1} = \{\text{good, fair, poor}\}$

$W_M = \{\text{RM, 1" overlay, 3"overlay}\}$

$W_{C2} = \{\text{good, fair, poor}\}$

Figure 8.   A valuation network for PMS decision making.

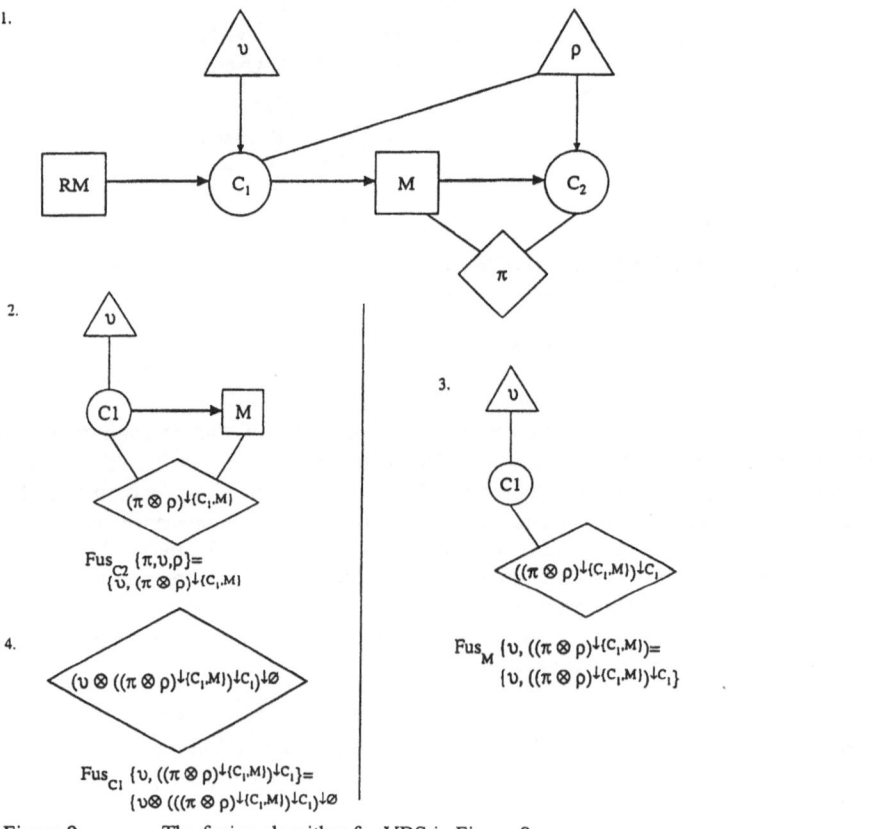

Figure 9.   The fusion algorithm for VBS in Figure 8.

5 CONCLUDING REMARKS

The paper presents a general overview of selected probabilistic graphical networks and their potential applications in pavement management decision making. These graphs capture both qualitative and quantitative formulations of the problem. The graphical probability networks are more efficient in handling uncertainty and other decision making problems than traditional approaches like decision trees which are extensively used in PMS.

ACKNOWLEDGEMENTS
The writers are deeply grateful to Adjo Amekudzi for her comments.

REFERENCES
Agogino, Alice M., et al. 1992. Decision-Analytic Methodology for Cost-Benefit Evaluation of Diagonostic Testers. *IIE Transactions.* 24 (1):39-54.
Amekudzi, A. & Attoh-Okine, N.O. 1997. Conceptual Frameworks for Understanding Brownfields Redevelopment Issues. *Infrastructure.* 2(2): 45-59.
Attoh-Okine, N.O. 1995. Pavement Management Decision Analysis Using Belief Functions in Valuation-Based Systems. *Third International Symposium on Uncertainty Modeling and Analysis*: 275-280. College Park: Maryland.
Attoh-Okine, N.O. 1997.Valuation Based Systems for Pavement Management Decision Making. *Chapter 11, in Uncertainty Analysis in Engineering and Sciences:Fuzzy Logic, Statistics and Neural Network Approach: Edited by Bilal Ayyub and Malan Gupta*: 157-170.
Attoh-Okine, N.O & Roddis Kim 1998. Uncertainties of Asphalt Layer Thickness Determination in Flexible Pavements-Influence Diagram Approach. *Çivil Eng and Env Systems*.15:107-124.
AASHTO Guidelines for Pavement Management Systems 1990. *AASHTO Guidelines.*Washington DC.
Clemen, Robert T 1991 Making Hard Decisions. *An Introduction to Decision Analysis.* PWS-Kent Publishing Company, Boston.
DPL. *Decision Analysis Software for Microsoft Windows* 1995. Student Edition. ADA Decision Systems.
Geiger, D. & Hackerman, D.1998. Systems and Humans Part A. *IEEE Transactions on Systems, Man and Cybernetics.* 28(1).
Greve, J.& Foulom, D. 1992. U-Valuation Networks for Bayesian Analysis. Working Paper Research Center, DK-8830 Tjele, Denmark.
Jensen, F.V 1996. An Introduction to Bayesian Network, Springer.
Kirkwood, Craig, W. 1992. An Overview of Methods for Applied Decision Analysis. *Interfaces.* 22(6):28-39.
Kulkarni, R.B. 1984. Dynamic Decision Model for a Pavement Management System. *Transportation Research Record* 997:11-18.
Maxwell, Daniel, T. 1996. Three Packages for Processing Influence Diagrams: DPL^{TM}, InDia^{TM} and Demos^{TM}. *Journal of Multi-Criteria Decision Analysis.* 5:72-74.
Shachter, Ross D. 1986. Evaluating Influence Diagrams. *Operations Research.* 6:871-882.
Shachter, Ross D. 1988. Probabilistic Inference and Influence Diagrams. *Operations Research.* 36(4):589-604.
Shafer, Glenn. 1992. Influence Diagrams and Decision Trees. *Note for Business* 876, University of Kansas.
Shenoy, P.P. 1991. Valuation Networks, Decision Trees, and Influence Diagrams: A Comparison, *School of Business, University of Kansas Working* Paper. 227.

Artificial Intelligence and Mathematical Methods in Pavement and Geomechanical Systems, Attoh-Okine (ed.)
© *1998 Balkema, Rotterdam, ISBN 90 5809 028 0*

A fuzzy logic based methodology to compare the level of security in public transportation agencies

A. Caballero
Department of Construction Management, Florida International University, Miami, Fla., USA

D. I. Ospina
Department of Civil and Environmental Engineering, Lehman Center for Transportation Research, Florida International University, Miami, Fla., USA

ABSTRACT: This paper presents a fuzzy logic based methodology to compare the security level of public transit systems using five of the most significant transit security-related indicators identified from different sources including a literature review, survey questionnaires, and telephone interviews with a number of public transportation agencies throughout the US. The paper includes a primary choosing of appropriate family of parameterized membership functions and information from experts to determine the parameters of the membership functions used for the analysis.

The methodology presented within this document might serve as a base for transit authorities and security experts to identify the agency, year, station/stop, etc. that had an acceptable security level. A close examination of the actual conditions will assist the transit and security authorities to determine how the security personnel, technologies and strategies can be most effectively used for maintaining a transit system that operates with the safest and most secure environment.

1 INTRODUCTION

The Federal Transit Administration's (FTA) Safety and Security Program has strongly encouraged transit officials to achieve the highest practical level of safety and security in all modes of transit. To meet this requirement, transit officials are currently implementing, or looking for, innovative and effective security plans and programs to enhance the security of passengers, transit employees, station properties, surrounding areas, and onboard vehicles. Transit agencies are convinced that high levels of security in the transit systems contribute to the confidence, reliability and comfort of customers, and therefore to the overall success in the operation of the systems.

The FTA published the Transit System Security Program Planning Guide and the Transit Security Procedure Guide reports (USDOT, 94) in 1994 as an aid for transit authorities to implement a program that maximizes the system security. Transit agencies had to report to the USDOT a complete System Security Program Standard in the first months of 1998.

Once the programs are implemented, transit authorities and security managers need to evaluate if the programs in place are effective in eliminating, mitigating and handling security threats and breaches. However, it has been found to be extremely complicated to evaluate and compare the level of security of the agencies before and after the implementation of the programs. Firstly, there was a lack of standard methods for collecting, reporting, and processing security-related data before the guidelines provided by the FTA. Secondly, there is not an accurate definition and a reliable methodology for determining the levels of security of a transit agency.

It was indicated by Zadeh in his publication entitled "Outline of a New Approach to the Analysis of Complex Systems and Decision Process" that, as the complexity of a system increases, our ability to make precise and yet significant statements about its behavior diminishes until a threshold is reached beyond which precision and significance become almost mutually exclusive characteristics (Zadeh, 1973). Using fuzzy logic provides the possibility of carrying out a more complete analysis and has major advantages when using fuzzy propositions for simplifying the adjustment of the knowledge representation in place of crisp rules.

2   MODEL DEFINITION

One of the most important tasks in the analysis and optimization of any process is the model definition. The obtained results will depend not only on how accurately the process is described in mathematical terms, but, more importantly, how to make proper assumptions and approximations. In this manner, the model may realistically characterize the system. The main tasks to be considered when defining a fuzzy model are as follows (Jang, 96):

1. Choose an appropriate family of parameterized membership functions.
2. Interview human experts familiar with the target system to determine the parameters of the membership functions used in the rule base.
3. Refine the parameters of the membership functions using regression and optimization techniques.

The development of the fuzzy model presented in this paper followed only tasks 1 and 2 and the three basic aspects for applying the model are introduced below:

2.1 Transit security-related indicators:

Number of crimes/year
Percent of arrests/crimes
Number of security personnel/year
Budget invested in security personnel/year
Budget invested in security equipment/year

2.2 Fuzzy set vocabulary (same fuzzy set vocabulary variables are used for all indicators):

Low
Moderate
Elevated
High

2.3 Domain of each variable:

It is the basic point for the model's correct behavior and for practical applications. The domain should be decided after a detailed study of the particular characteristics of the agencies. Figures 1 through 5 show a set of functions defined as a first approximation using values obtained from nine security and transit experts from three selected transit agencies operating bus systems (information from the three experts at each transit agency was averaged for the analysis).

Additionally, an accurate definition of the α-*Cut Threshold* is important when defining the model because it permits a determination of whether or not an indicator will actively participate in the calculation. In this particular study, α-*Cut Threshold* was selected as 0.1.

The method of composition to be used is the average of a membership in which the Compatibility Index (CI) for each agency, related to some specified situation, is calculated by means of [Cox, 95]:

$$C I = [\sum_{i=1}^{N} \mu_I(x)] / N \qquad \qquad (i)$$

where $\mu_I(x)$ is the degree of membership for the indicator in the agency; and N, the number of indicators.

3 A COMPARATIVE ANALYSIS FOR THE THREE PARTICIPATING AGENCIES

The proposed Fuzzy logic based methodology to compare the security level of public transit systems is illustrated in this paper using data obtained from the research project entitled "Analysis of Technologies and Methodologies Adopted by US Transit Agencies to Enhance Transit Security" (Shen, 97). A number of transit agencies operating in a similar environment and with similar system characteristics were identified from the report as candidates for the application of the proposed methodology. The transit mode selected for the study was the bus system; the most common transit mode operated by all transit agencies.

The comparison was performed among three of the participating agencies that were identified as agencies A, B, and C for this particular study. The selection was made based on the accuracy and completeness of the information. The same procedure could be followed to compare the security level of one agency during different years.

Table 1:   Bus System's Indicators for Agencies A, B, and C

Agency	No. of Crimes	No. of Arrests	No. of Sec. Personnel	Budget Sec. Personnel	Budget Sec Equipment
A	1400	40	30	$1,600,000	$45,000
B	2200	150	50	$1,900000	$40,000
C	1600	90	45	$2,300,000	$60,000

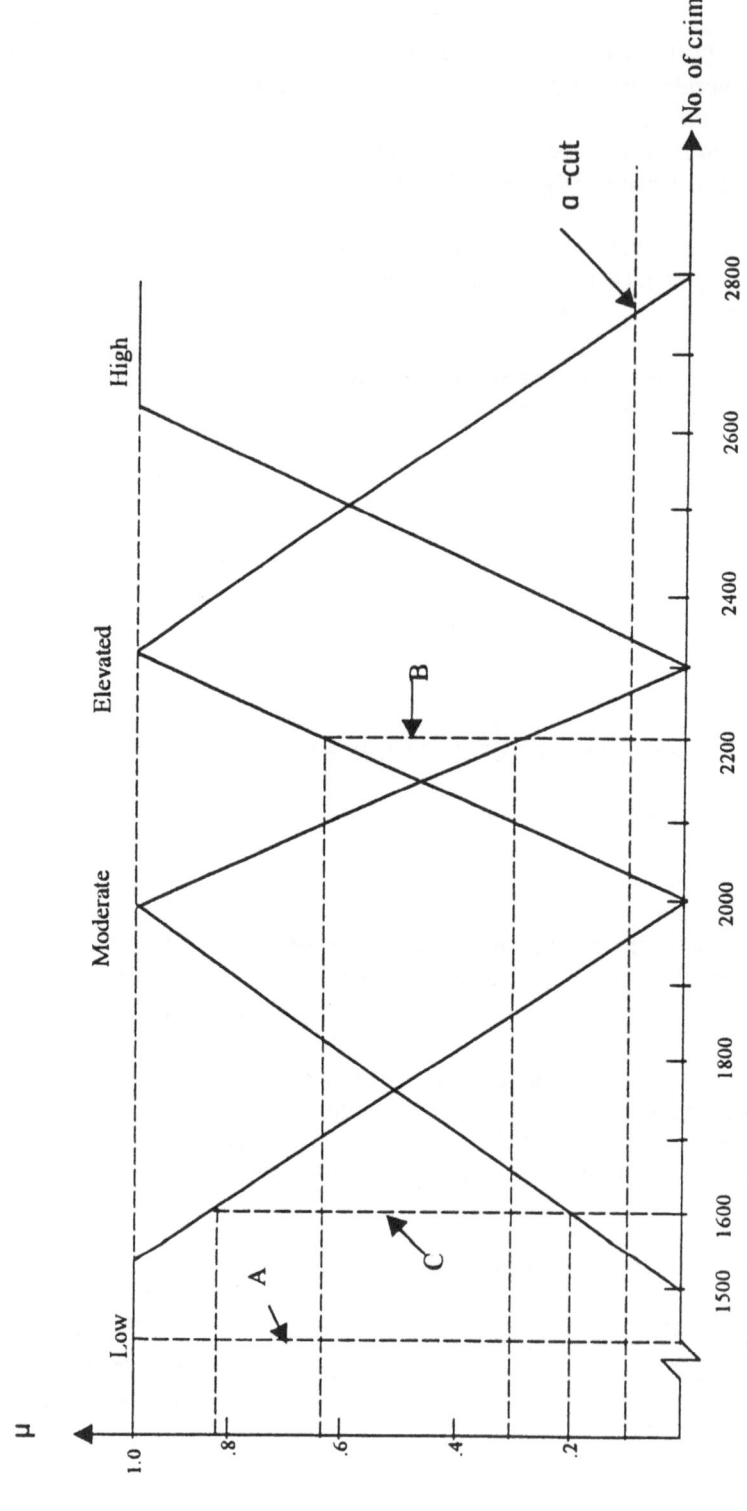

Fig.1 Fuzzy set for Number of Crimes

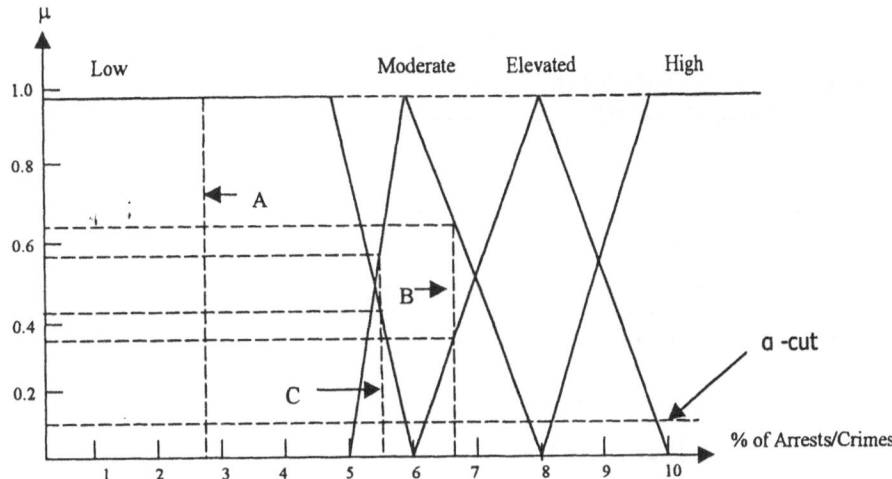

Fig. 2 Fuzzy set for Percent of Arrests/Crimes

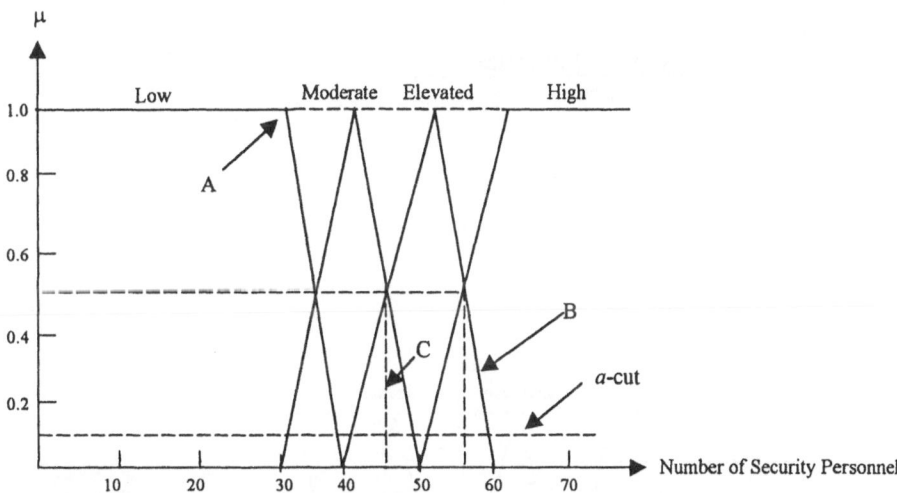

Fig. 3 Fuzzy set for Number of Security Personnel

In the example, it was accepted that the domain of each variable fulfills the values as it is indicated in Figures 1 through 5. A query is obtained using fuzzy sets representing the grade a particular agency approaches some selected conditions. When a fuzzy model is applied, it is possible to reduce the difference between the way an expert visualizes a problem and its solution by encoding knowledge at a higher level. The obtained results give more accurate information related to the underlying meaning than using crisp lines.

This situation can be presented, as an example, in the following way: It is required to compare different agencies with the following characteristics:

No. of Crimes was **Moderate,**
Percent of Arrests/crimes were **Low,**
No. of Security Personnel was **Moderate,**

139

Budget invested in Security Personnel was **Elevate,** and
Budget invested in security equipment was **Elevate**, then
Using figures 1 through 5 and replacing the obtained results in Formulae 1 as follows:

Agency A:
CI = [0+1.0+0+0+1.0] / 5
CI = 0.40

Agency B:
CI = [0.60+0+0+0+0.20] / 5
CI = 0.16

Agency C:
CI = [0.19+0.41+0.5+0.5+0] / 5
CI = 0.32

The obtained results are shown in the following query:

Agency	Compatibility Index
A	0.40
C	0.32
B	0.16

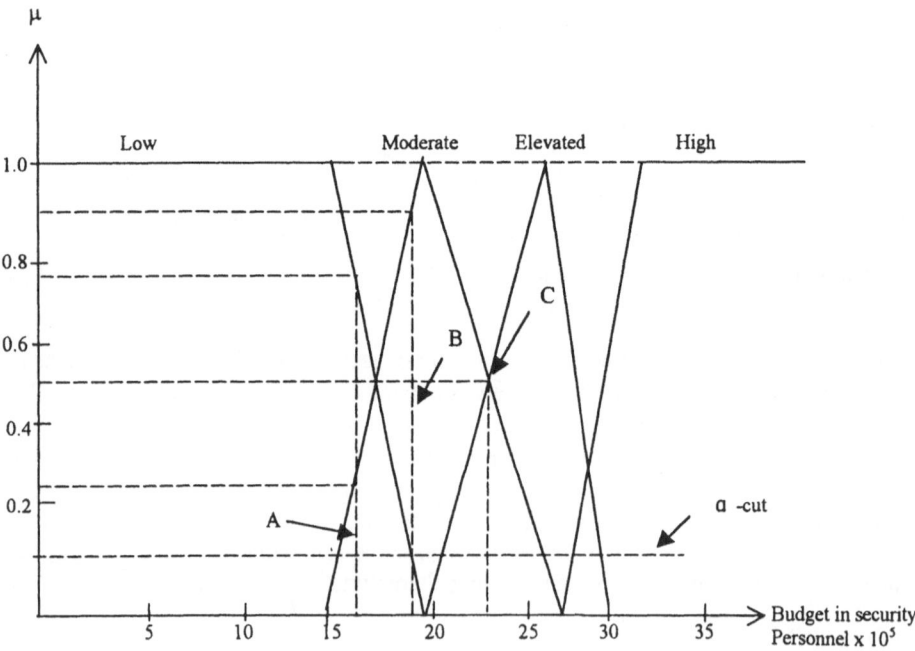

Fig. 4 Fuzzy set for Budget in security Personnel

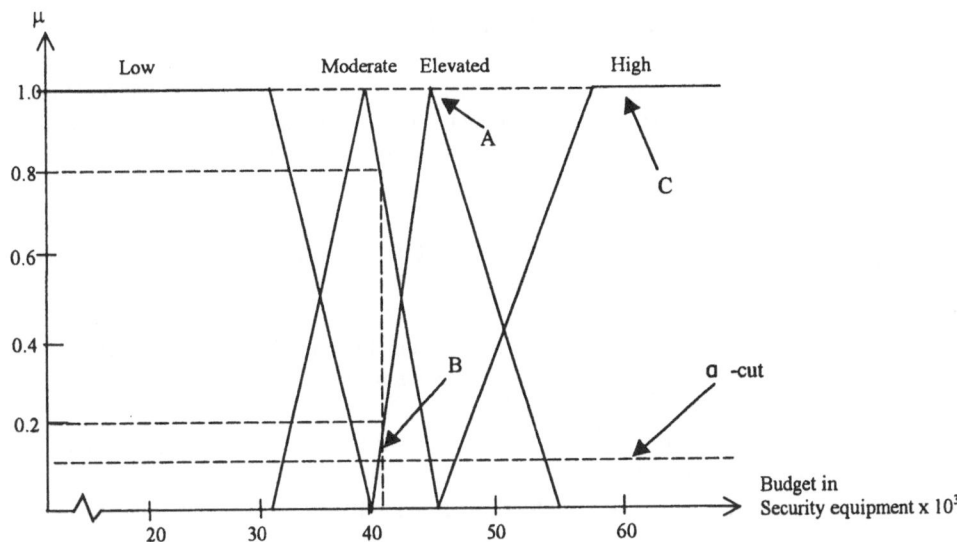

Fig. 5 Fuzzy Set for Budget in Security Equipment

The compatibility index gives an idea of how much each agency approaches the established conditions. In this comparison, agency A was the one that best observed the presented situation. This does not mean that this is the agency with the best results, because the comparison can be made using different criteria depending on the needs or the expert's preferences once the evaluation is carried out.

It is possible, for example, to define the optimum solution or a more realistic one and obtain a query that gives the order in which it fits each agency. Lets assume that the conditions in the previous example are all taken as **Moderate**. In this hypothetical situation, the query will be:

Agency	Compatibility Index
B	0.58
C	0.32
A	0.04

The order in the query changed with respect to the order obtained before. The situation that better characterizes each agency can also be obtained using the compatibility index as follows:

Indicator	Agency A	Agency B	Agency C
No. of Crimes	Low	Elevate	Low
Percent of Arrests/crimes	Low	Moderate	Moderate
No. of Security Personnel	Low	Elev./High	Mod./Elev
Budget invested in Security Personnel	Low	Moderate	Mod./Elev
Budget invested in Security Equipment	Elevate	Moderate	High
Compatibility Index	0.96	0.69	0.68

4 CONCLUSIONS AND RECOMMENDATIONS

1. The proposed method possesses the flexibility and possibility of knowledge representation that characterizes the fuzzy models. These models are adequate for solving difficult, computational, complex and imprecise situations as it exists when evaluating the security level in transportation agencies.

2. This paper is limited to the presentation of a solution method. In order to accept the obtained results as certain, a more detailed study is necessary that includes at least the following:
   a) Definition of fuzzy sets and their domain for each indicator should be discussed with transit and security experts from additional agencies in order to obtained more precise results.
   b) To establish a comparison of the agencies using standard units, such as number of annual passenger miles per directional mile, annual unlinked passenger trips per vehicle revenue mile, etc. and not the averaged totals as taken in the example. This is because the size of the agency should not affect the comparative analysis.
   c) A more precise definition of the indicators to take into account, which could include aspects, that were not considered in this work.

3. The presented method may be used as a base for the development of an easy-to-use computer software that will allow the transit and security authorities to measure the security level of the agencies, modes, routes, and/or facilities. This measure will make transit authorities aware of the condition of system security as compared to other systems in the nation and might serve as a guide in the process of achieving the highest practical level of transit security as encouraged by the FTA.

REFERENCES

Cox, E.D. 1995. *Fuzzy Logic for Business and Industry*. Rockland: Charles River Media.

Jang J-S.R., C.-T. Sun, E. Mizutani 1996. *Neuro-Fuzzy and Soft Computing: A Computational Approach to Learning and Machine Intelligence*. Prentice Hall.

Kosko, B. 1997. *Fuzzy Engineering*. New Jersey. Prentice Hall.

New Jersey Transit Police Department, 1998. Security in Public Transportation Agencies Data Sets.

Shen L.D., Ospina D., Zhao F., Elbadrawi, H., 1997. *Analyses of Technologies and Methodologies Adopted by US Transit Agencies to Enhance Transit Security*. LCTR, Research Findings.

U.S. Department of Transportation, Research and Special Programs Administration, Volpe National Transportation Systems, 1994. *Transit System Security Program Planning Guide*. FTA-MA-90-7001-94-1, DOT-VNTSC-FTA-94-1.

U.S. Department of Transportation, Research and Special Programs Administration, Volpe National Transportation Systems, 1994. *Transit Security Procedures Guide*. FTA-MA-90-7001-94-2, DOT-VNTSC-FTA-94-8.

Zadeh, L.A. 1973. *Outline of a New Approach to the Analysis of Complex Systems and Decision Process*. IEEE Transactions on Systems, Man, and Cybernetics, January, 1973, pp. 28-44

Artificial Intelligence and Mathematical Methods in Pavement and Geomechanical Systems, Attoh-Okine (ed.)
© *1998 Balkema, Rotterdam, ISBN 90 5809 028 0*

Identification of landfill settlement characteristics by fuzzy analysis of interactive systems dynamics

B.Tansel
Drinking Water Research Center and Civil and Environmental Engineering Department, Florida International University, Miami, Fla., USA

ABSTRACT: The dynamic processes taking place in sanitary landfills are very complicated because of the interactive nature of physical, chemical, and biological changes taking place. Modeling of a landfill is further complicated by the heterogeneity of the deposited waste materials, differences between the operational practices among the landfills, and geographical and seasonal nature of the waste characteristics as well as operational practices. A fuzzy modeling approach for identification of settlement characteristics can be successfully developed since a significant knowledge based exists about the operation and performance of specific landfills. In this paper a surface structure was developed for use to identify settlement characteristics of landfills by incorporating interactive system dynamics into the definition of fuzzy system parameters.

1 INTRODUCTION

Sanitary landfill is an engineering project that requires detailed planning and specifications, careful construction, and efficient operation. In a landfill solid wastes are disposed by spreading them in thin layers, compacting to smallest practical volume, and covering with earth each day in a way that minimizes environmental problems. After solid wastes are deposited in a landfill, physical, chemical and biological processes take place, as wastes are decomposed (Diaz et al. 1982). Settlement of landfills is an important planning and design criterion for defining the end use of a closed landfill. Although closed sanitary landfill sites have been developed for use as open areas for recreation and industrial and commercial areas with structures, the extent of settlement has been an important factor in the success of the functional use of the closed landfill sites. A number of methods have been used to increase the bearing capacity and to reduce the extent of settlement of landfill. Table 1 presents a summary of the ground improvement methods which have been utilized in landfills before closure.

In general, landfill settlement can be identified as primary and secondary settlement. Initially, the primary settlement occurs rapidly then slows down. The settlement of a landfill continues after the primary compression. The long-term settlement is related to the biological decomposition rate of the deposited waste materials. About 90 percent of the settlement typically occurs in the first 5 years after the waste is deposited, although it may continue for 25 or more years at a slower rate (Conrad et al., 81). Factors that affect the extent and rate of subsidence include (Yen and Scanlon 1975, Charles 1984, Rao et al. 1977, Zimmerman et al. 1977, Lutton et al. 1979, Merz and Stone 1962):

1. Extent of compaction
2. Waste characteristics (composition organic content, heterogeneity and biodegradability)

Table 1. Ground improvement techniques to reduce minimize extent and rate of settlement in landfills before closure.

Technique	Procedure	Criteria	References
Increased cap thickness	Soil thickness over the landfill is increased.	The thickness of the soil should be at least 1.5 times the width of structural footing.	Sowers 1968 Lutton et al. 1979
Increased cover-to-waste ratio	Amount of daily cover is increased.	For efficient utilization of landfill space, cover-to- waste ratio should be increased only at the areas below the intended foundations.	Brunner and Keller 1972
Increased compaction	In-place refuse density is increased.	Refuse density can be increased by using effective compaction equipment or by landfilling techniques that provide increased in-place densities such as baling or shredding.	Lutton et al. 1979
Additives and cements	Incorporation of stabilizers into the soil during cover placement can improve the bearing capacity.	Common materials mixed with soil include lime, portland cement, and organic chemicals. Wetting, mixing, compacting and curing are critical steps. Achieving a homogeneous mixture is difficult.	Lambe and Whitman 1979 Lutton et al. 1979
Selective disposal	The waste suitable for foundations is placed in the designated areas for future construction.	The waste suitable for foundations can be placed and compacted in the areas to support planned buildings, e.g., fly ash and bottom ash can be placed at the assigned areas.	Lutton et al. 1979
Undisturbed structural pads	Pedestals of undisturbed ground are left as structural pads.	Future settlement problems are eliminated at the areas of construction.	Lutton et al. 1979
Decomposition rate acceleration	Rate of decomposition is increased during the landfill operation.	Refuse decomposition can be accelerated by various techniques such as leachate recycling, and wetting; or refuse can be partially composted before landfilling.	Lutton et al. 1979

3. Age of fill
4. Depth of fill
5. Climate
6. Amount and type of soil cover
7. Water table

A landfill with high percentage of demolition and construction waste is expected to settle less compared with a fill with high organic content. The amount of water in the fill significantly affects the decomposition rate. A landfill with limited available water for biochemical decomposition will settle slowly. For example, in Seattle, where rainfall exceeds 30 inch per year, a 20 feet fill had settled 4 feet during the first year after completion (Dunn 1957). In Los Angeles, where less than 15 inch of rain falls per year, a 75 feet high area had settled only 2.3 feet and a 46 feet high area had settled only 1.3 feet in three years after closure (County of Los Angeles 1973).

The composition of solid waste within a particular section of a landfill depends on the season of deposition and the origin of the solid wastes (i.e., industrial or commercial). In general, 40 to 80 percent of the refuse is biodegradable. Bulky metallic objects create large voids in the landfill when, due to corrosive action within the medium, they are weakened to the point which they collapse under the above loads. This collapse can result in large, local depressions in the ground surface.

The in-place density of the solid waste fills varies significantly ranging from 800 to 1,600 pounds per cubic yard depending on the type of the compaction equipment and fill technique used (baling, shredding or piling). In-place density of a landfill influences other characteristics of the landfill such as bearing capacity, decomposition rate of solid waste, and extent and rate of settlement (Brunner and Keller 1972).

The landfill behavior could be represented in a matrix form based on the available information, observations, experiences, and public records. Although some of the information is valuable, may not be usable by a mathematical model since it is not easily quantifiable in terms of mathematical parameters and/or the information may not be scientifically collected (i.e., personal observation). However, without extensive modeling effort, one can develop a sample knowledge base (i.e., domain knowledge) as shown in Table 2. These types of discrete information sources could be utilized in a fuzzy approach to estimate the settlement characteristics of a landfill.

Development of a model for estimating the extent and rate of settlement in a landfill is extremely complex. Thus, when developing a model to predict extent and rate of settlement; waste characteristics, climate information, as well as landfill development/operational practices should be considered. The uncertainty associated with some of the input parameters complicates the structure of the model. The fuzzy set theory is an extremely convenient mathematical tool which can be easily adapted for use to solve problems with uncertainty, subjectivity, ambiguity and indetermination (Teodovic et al. 1994). This paper develops a model to estimate extent and rate of settlement in landfills. The rule base was developed in relation to landfill dynamics and significance of the dynamic phenomena in relation to the age of the landfill.

2 FUZZY ANALYSIS FRAMEWORK

The fuzzy inference system is a computing framework based on the concepts of fuzzy set theory, fuzzy if-then rules, and fuzzy reasoning. The basic structure of the fuzzy inference system consists of three conceptual components:

-     rule base
-     data base (or dictionary)
-     reasoning mechanism

In general, the fuzzy inference system is designed based on the past known behavior of a target system. Fuzzy modeling approach has the following advantages (Jang et al. 1997):

1.   The rule structure of a fuzzy inference system makes it easy to incorporate human expertise about the target system directly into the modeling process. Fuzzy modeling utilizes the domain knowledge that might not be easily or directly employed in other modeling processes.

2.   When the input-output data of a target system is available, conceptional system identification techniques can be used for fuzzy modeling. In other words, the use of numerical data also plays an important role in fuzzy modeling.

Fuzzy modeling can be performed in two stages: 1. surface structure, and 2. deep structure. Identification of the surface structure involves selection of relevant input and output variables, selection of the specific fuzzy inference system, determination of the number of linguistic terms associate with each input and output variables, and designing of a collection of fuzzy if-then rules.

Identification of the deep structure involves selection of an appropriate family of membership functions (MF), interview of human experts familiar with the system, and refining the parameters of the MF using regression and optimization techniques.

3 DEVELOPMENT OF SURFACE STRUCTURE

Although the landfill dynamics is complex, based on the domain knowledge, only two of the dynamic processes have direct impact on landfill settlement: moisture transport and biodecomposition. These two processes constitute the major components of the fuzzy inference system to be correlated to settlement process. The relevant input variables for the moisture transport and biodecomposition and the relevant output variables for the settlement are summarized in Table 3.

Table 2. Sample knowledge base for landfill dynamics.

	0-1 yr	3 yrs	10 yrs	30yrs
Fluid flow				
Unsaturated	x	x		
Saturated				
Biodecomposition				
Slow	x			
Intermediate				
Steady				
Gas production				
Slow	x			
Intermediate		x		
Steady				
Solute transport (Leachate quality)				
Weak	x			
Intermediate				
Strong				
Steady				
Ambient temperature				
Warm-Seasonal	x			
Moderate-Seasonal				
Cold-seasonal				
Tropical				
Precipitation				
Wet	x			
Moderate				
Dry				
Deformation				
Primary	x			
Transition				
Secondary				

Table 3. Input variables for the moisture transport and biodecomposition and outputs for settlement characterization.

Dynamic Condition	Input variable	Parameter code (linguistic terms)
Water transport	Waste in-place density	densely compacted medium compacted lightly compacted uncompacted
	Waste composition	MSW with high food waste municipal solid waste (MSW) MSW with high C&D waste construction & demolition waste (C&D)
	Cover/waste ratio	high average low none
	Precipitation	heavy medium light dry
	Depth of fill	very deep deep average shallow
Biodegradation	Waste in-place density	densely compacted medium compacted lightly compacted uncompacted
	Composition	MSW with high food waste municipal solid waste (MSW) MSW with high C&D waste construction & demolition waste (C&D)
	Cover/waste ratio	high average low none
	Ambient temperature	tropical warm-seasonal average-seasonal cold-seasonal
	Precipitation	heavy medium light dry
	Age of fill	new fairly new fairly old old
Settlement	Cracks	none new small surficial new large surficial old small surficial old large surficial
	Depressions	none new small new medium new large old small old medium old large

Based on the experience, a number of rules can be developed for settlement behavior of landfills. For example, for construction and demolition landfills, the settlement would be minimal since biodecomposition would be insignificant. The insignificant parameters can be eliminated based on the available knowledge base from the past experiences.

4 APPLICATION OF THE FUZZY MODEL

The extent and rate of settlement can be represented in terms of size and age of cracks and depressions forming on the landfill surface. The settlement can be represented as a matrix in terms of crack and depressions as follows:

$$
S = \begin{array}{c} \\ c \\ \downarrow \\ 1 \\ 2 \\ 3 \\ 4 \\ 5 \end{array}
\begin{array}{ccccccc}
d \rightarrow \ 1 & 2 & 3 & 4 & 5 & 6 & 7 \\
\left(\begin{array}{ccccccc}
S_{11} & 0 & 0 & 0 & 0 & 0 & 0 \\
S_{21} & S_{22} & S_{23} & S_{24} & 0 & 0 & 0 \\
S_{31} & S_{32} & S_{33} & S_{34} & 0 & 0 & 0 \\
S_{41} & S_{42} & S_{42} & S_{44} & S_{45} & S_{46} & S_{47} \\
S_{51} & S_{52} & S_{53} & S_{54} & S_{55} & S_{56} & S_{57}
\end{array} \right)
\end{array}
$$

where:

S_{ij} = type of crack and type of depression observed on the landfill surface

and the rows and columns are defined based on:

c = crack types
d = depression types

Based on the dynamic phenomena taking place in a landfill, two indices can be defined:

m_{rs} = moisture transport index
b_{rs} = biodegradation index

These indices are defined as follows:

m_{rs} = f (waste in-place density, waste composition, cover/waste ratio, precipitation, depth of fill)

b_{rs} = f (waste in-place density, waste composition, cover/waste ratio, ambient temperature, precipitation, age of fill)

The larger index represents a more significant settlement. The settlement can be defined as:

$$S_{ij} = \alpha\, m_{rs} + \beta\, b_{rs}$$

where:

α = biodegradation parameter (i.e., secondary settlement), numerical value 0-1
β = moisture parameter (i.e., primary settlement), numerical value 0-1

Table 4. Parameter identification values for the algorithm.

Settlement	Waste in-place density (lb/cu yd)	Waste composition (based on biodegradable fraction, 4 = most 1 = least biodegradable)	Cover-to-waste ratio	Precipitation (in/yr)	Ambient temperature (4 = most favorable for biodegrad-ation, 1 = least favorable for biodegrad-ation)	Depth of fill (ft)	Age of fill (yrs)
c1-c5 d1-d7	> 1000 800-1000 400-600 < 400	4 3 2 1	1/2 1/2-1/3 1/4 0	> 50 40-50 20-40 < 20	4 3 2 1	> 50 30-50 15-30 < 15	> 30 20-30 10-20 0-10

4^7 = 16384 different input combinations = = = > for 35 outputs for characterizing settlement.

The parameters α and β are defined in the rules of approximate reasoning algorithm. For example, for newly closed landfills, α will be zero. During the first five years after the closure β will be larger than α, while α increases linearly during this period. The values of α are assigned according to knowledge base about the biodegradation trends in landfills. Similarly, the values of β are assigned according to knowledge base about the saturation (i.e., time to reach field capacity) trends in landfills.

4.1 *Approximate reasoning algorithm*

The approximate reasoning algorithm determines the moisture transport and biodegradation indices. Table 4 presents the parameter identification values for the algorithm.

Rule 1: If waste in-place density is "densely compacted," then the m_{rs} is 1
else
Rule 2: If waste in-place density is "medium compacted," then the m_{rs} is 2
else

.
.
.

Rule n: If waste in-place density of "densely compacted," then b_{rs} is 1

.
.
.

By applying the approximate reasoning algorithm, settlement characteristics are obtained. A sample output from the program is presented in Table 3.

5 CONCLUSIONS

The rate and extent of settlement in landfills is one of the most critical factors that must be considered when planning a final site use for closed landfills. Due to the heterogeneity of the composition of solid waste based on seasons and geographic location; differences among the regional operational guidelines, policies, and regulations; and lack of standardized measurement guidelines for landfill operation and performance after closure; mathematical modeling of landfill settlement landfills have not been successfully attempted. The knowledge based, although subjective, exists about the performance and operation of landfills. A fuzzy system identification approach could be utilized to characterize landfills. The coupling of fuzzy parameters with the actual dynamic processes could provide significant information for predicting the extent and rate of settlement for a landfill.

REFERENCES

Brunner, D.R. & D.J. Keller 1972. *Sanitary landfill design and operation*. U.S. Environmental Protection Agency, SW-65ts.

Charles, J.A. 1984. Settlement of fill. In: *Ground movements and their effects on structures*. Edited by P.B. Attewell and R.K. Taylor. Surrey University Press. London, UK.

Conrad, E.T., J.J. Walsh, J. Atcheson & R.B. Gardner 1981. *Solid waste landfill design and operation practices*. EPA Draft Report, Contract No. 68-01-3915.

County of Los Angeles 1973. *Development of construction and use criteria for sanitary landfills*. California, Dept of County Engineer. Final Report on Solid Waste Management Demonstration Grant. PB 218 672, EPA-SW-D-73.

Diaz, L.F., G.M. Savage & C.G. Golueke 1982. *Resource recovery from municipal solid waste*. Volume II, Final Processing, CRC Press.

Dunn, W.L. 1957. Settlement and temperature of a covered refuse dump. *Trend in engineering*. University of Washington, 9(1):19-21.

Lambe, T.W. & R.V. Whitman 1979. *Soil mechanics*. Wiley, New York.

Lutton R.J., G.L. Regan & L.W. Jones 1979. *Design and construction of covers for solid waste landfills*. Report for U.S. Environmental Protection Agency, EPA-600/2-79-165.

Jang, J.-S.R., C.-T. Sun & E. Mizutani 1997. *Neuro-fuzzy and soft computing: A computational approach to learning and machine intelligence*. Prentice Hall Inc. Upper saddle River, New Jersey.

Merz, R.C. & R. Stone 1962. Landfill Settlement Rates. *Public Works*. 93(9):103-106,210-212.

Oweis, I.S. & R.P. Khera 1998. *Geotechnology of waste management*. Second edition. PWS Publishing Co. Boston, Massachusetts.

Rao, S.K., L.K. Moulton & R.K. Seals 1977. Settlement of refuse landfills. *Proc. The Conference on Geotechnical Practice For Disposal of Solid Waste Materials*. June 13-15. pp. 574-598. The University of Michigan Specialty Conf. Geotechnical Division, ASCE. Ann Arbor, Michigan.

Sowers, G.F. 1968. Foundation problems in sanitary landfills. *J. Sanitary Eng. Div. ASCE*. 94(SA1):103-116.

Teodorovic, D., M. Kalic and G. Pavkovic 1994. Potential for using fuzzy set theory in airline network design. *Transpn. Res. Bull.* 28B(2):103-121.

Yen, B. C., and B. Scanlon 1975. Sanitary landfill settlement rates. *J. Geotechnical Eng. Div. ASCE*. 101(GT5):475-487.

Zimmerman, R., W.H. Chen & A.G. Franklin 1977. Mathematical Model For Solid Waste Settlement. *Proc. The Conference on Geotechnical Practice For Disposal of Solid Waste Materials*. June 13-15. pp. 210-226. The University of Michigan Specialty Conf. Geotechnical Division, ASCE. Ann Arbor, Michigan.

Artificial Intelligence and Mathematical Methods in Pavement and Geomechanical Systems, Attoh-Okine (ed.)
© *1998 Balkema, Rotterdam, ISBN 90 5809 028 0*

Predicting the performance of surfactant-enhanced subsurface remediation

Shonali Laha
CEE/DWRC, Florida International University, Miami, Fla., USA

Sumitra Mukherjee
SCIS, Nova Southeastern University, Ford Lauderdale, USA

ABSTRACT: Enhanced pump-and-treat methods have received considerable attention in recent times as more traditional methods for groundwater remediation prove ineffective for target contaminants. The addition of aqueous surfactant solutions to the subsurface have produced increased solubilization of hydrophobic compounds and improved efficiencies of soil washing/flushing applications. In order to predict the performance effectiveness of surfactant-enhanced remediation it is necessary to obtain information on compound hydrophobicity, surfactant properties, and aquifer characteristics. These data are derived from field measurements and laboratory treatability experiments. This paper examines the use of nonlinear mixed integer programming techniques and neural networks at predicting the surfactant solubilization of hydrophobic contaminants.

1 INTRODUCTION

Subsurface contamination by hydrophobic organic compounds (HOCs) and nonaqueous phase liquids (NAPLs) is a matter of growing concern because groundwater is a limited resource. Such contamination results from accidental spills, inadequate waste disposal procedures, and poor operational practices at commercial/industrial facilities. Conventional pump-and-treat systems are generally ineffective in remediating sites contaminated with HOCs and NAPLs because of the limited aqueous solubilities of these chemicals and their tendency to remain associated with the soil structure.

For the treatment of HOCs and NAPLs it has been suggested that the performance of pump-and-treat methods may be enhanced by the addition of HOC/NAPL-solubilizing agents such as commercially available surfactants. Surfactants are amphiphilic molecules possessing both hydrophilic and hydrophobic parts. When present in aqueous solution, surfactant molecules align themselves along interfaces resulting in decreased interfacial tension. At concentrations exceeding critical micelle concentration (cmc), surfactant molecules self-aggregate into colloidal-sized clusters known as micelles. Surfactant molecules in a micelle are oriented with the hydrophilic head facing outward into the aqueous phase and the hydrophobic tail constituting a hydrocarbon-like micellar core. Surfactant solubilization of HOCs is facilitated by the incorporation of HOCs into surfactant micelles, and enhanced HOC solubilities are generally observed at surfactant concentrations greater than cmc. The HOC "solubility" resulting from surfactant addition represents a pseudo-aqueous phase HOC concentration consisting of both the aqueous and micellar phase HOC content. Increased pseudo-aqueous phase HOC concentrations permit more effective removal of HOCs from the subsurface in soil washing and soil flushing methods commonly used for site remediation.

The presence of soils results in a more involved HOC partitioning process. HOCs may exist in the aqueous, micellar, and soil-sorbed phases. In addition, surfactants may also sorb onto soil, and HOCs may be associated with surfactant monomers to some degree. The theory of HOC partitioning in the context of a typical soil-water-surfactant system is discussed in the following section.

2 CONCEPTUAL MODEL OF HOC PARTITIONING

Solubility enhancement of HOCs by surfactants is governed by the partitioning of HOCs into surfactant micelles (or surfactant emulsions) present in the pseudo-aqueous phase. The solubility enhancement in aqueous surfactant systems has been represented as the ratio of apparent aqueous HOC solubility (S_w*) to the intrinsic solubility in surfactant-free water (S_w):

$$\frac{S_w^*}{S_w} = 1 + K_{mn} X_{mn} + K_{mc} X_{mc} \tag{1}$$

where K_{mn} and K_{mc} are the HOC partition coefficients between surfactant monomers and water and between surfactant micelles and water, respectively; X_{mn} and X_{mc} are the fractional concentrations of surfactant monomers and micelles, respectively. Surfactant concentration plays a critical role in the surfactant-enhanced solubilization of HOCs. In general, significant HOC solubilization is observed only at surfactant concentrations greater than cmc.

In the presence of soil, the HOC partitioning process is more complicated. HOC sorption in surfactant-free soil-water systems is believed to be dominated by HOC partitioning into the soil organic matter phase, and has been related to HOC hydrophobicity as follows (Karickhoff *et al.* 1979):

$$K_p = 0.63 f_{oc} K_{ow} \tag{2}$$

where K_p is a measure of the partitioning of HOC between soil and water, i.e., K_p represents the ratio of soil-sorbed HOC concentration (in mg/kg) and the aqueous phase HOC concentration (in mg/L); f_{oc} is the fraction organic carbon content of the soil sorbent; and K_{ow} is the dimensionless octanol-water partition coefficient (defined as the ratio of HOC concentration in octanol to the aqueous HOC concentration at equilibrium). K_{ow} represents the hydrophobicity of the partitioning solute or HOC: a high value for K_{ow} denotes a greater hydrophobicity, and generally suggests a lower aqueous solubility and greater tendency to partition into available organic phases. Several empirical models have been proposed for various classes of HOCs to predict HOC partitioning between soil and aqueous phases, and the organic carbon-normalized partition coefficient has been defined to facilitate projection of HOC phase partitioning (Karickhoff *et al.* 1979, Chiou *et al.* 1998):

$$K_{oc} = \frac{K_p}{f_{oc}} = 0.63 K_{ow} \tag{3}$$

However, in the presence of surfactant, additional competitive processes affect the distribution of HOCs, in particular:
- partitioning of HOCs into the micellar phase, and
- sorption of HOCs by soil-sorbed surfactants.

Depending on the net effect of these two processes, the apparent HOC soil-water distribution coefficient K_p* in systems containing surfactant may increase or decrease relative to the surfactant-free soil-water partition coefficient K_p. It is important to emphasize that all the partition coefficients described relate to equilibrium partitioning, and that the kinetics of the partitioning process are not considered. If the addition of surfactant to a soil-water system increases the apparent aqueous phase concentration (S_w^*), the apparent soil-water partition coefficient (K_p^*) is lowered; whereas a net decrease in aqueous HOC concentration results in greater K_p^*. The decrease in K_p^* on surfactant addition is more dramatic for the more hydrophobic compounds. Sun *et al.* (1995) report that K_p decreased from 478 to 2.36 at a particular surfactant dose for PCB ($S_w = 0.001$ mg/L); at the same surfactant concentration the K_p for the less hydrophobic phenanthrene ($S_w = 1.6$ mg/L) decreased only slightly from 13.3 to 6.02, and for the fairly soluble naphthalene ($S_w = 31.7$ mg/L) a slight increase in K_p^* was observed. The surfactant used in that study was an emulsion-forming surfactant. Similarly in another study using a micelle-forming surfactant Triton X-100, K_p^* values for DDT and PCB decreased significantly at surfactant concentrations exceeding cmc whereas for 1,2,4-TCB, a relatively water-soluble solute, the amount of solute sorbed increased yielding higher values for K_p^* (Sun *et al.* 1995).

The effect of a micelle-forming surfactant on HOC partitioning between soil and water has been expressed as (Sun *et al.* 1995):

$$K_p^* = K \frac{1 + X_{s/om} K_{s/om}}{1 + X_{mn} K_{mn} + X_{mc} K_{mc}} \tag{4}$$

where $K_{s/oc}$ is the HOC partition coefficient between sorbed surfactant and soil organic matter, and represents the effectiveness of sorbed surfactant as a sorption medium relative to the native soil organic carbon (i.e., $K_{s/oc} = K_s/K_{oc}$, where K_s is the partition coefficient of given HOC between sorbed surfactant and water). $X_{s/oc}$ is the fractional concentration of sorbed surfactant per unit mass of soil organic carbon. K_{mn}, K_{mc}, X_{mn}, and X_{mc} are as defined earlier.

3 PARAMETER ESTIMATION FROM LABORATORY DATA

Figure 1 (adapted from Laha 1992) illustrates typical solubilization data obtained for phenanthrene in soil-water systems using six different commercially available surfactants (listed

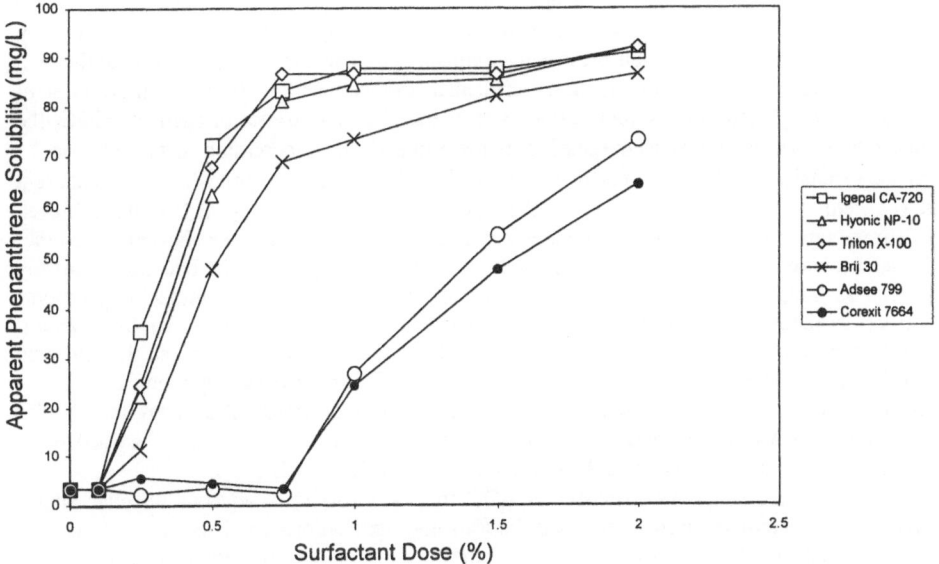

Figure 1. Surfactant Solubilization of Phenanthrene

to the right of figure). Phenanthrene is a polycyclic aromatic hydrocarbon (PAH) consisting of three fused benzene rings. PAHs are neutral hydrophobic HOCs and several have been listed as priority pollutants. The chemical structure of phenanthrene follows:

Phenanthrene ($C_{14}H_{10}$, molecular weight 178) is sparingly soluble in water with an aqueous solubility (S_w) reported at approximately 1.3 mg/L. It is very hydrophobic as indicated by a relatively high K_{ow} value (log K_{ow} = 4.56), and when present in soil-water environments demonstrates a tendency to sorb strongly onto soil. The addition of surfactants is, therefore, expected to result in greater apparent solubility and decreased partitioning onto soil organic matter. The data presented in Figure 1 were obtained from laboratory batch experiments employing soil-water systems at a soil-to-water ratio of 1 g: 8 mL. Surfactant dose is expressed as a percentage by weight with 1% representing approximately 10,000 mg/L of surfactant. Solubilization data is shown as apparent aqueous solubility in mg/L. Each system received the same amount of phenanthrene.

In the absence of surfactant, the pseudo-aqueous phase phenanthrene concentration is near reported aqueous solubility. As the surfactant dose increases, more phenanthrene is solubilized, resulting in phenanthrene solubilities $>10\times$ aqueous solubility (~60 to 80% of added phenanthrene is solubilized at 2% surfactant dose). It is apparent from Figure 1 that little solubilization occurs below a specific surfactant concentration (i.e., 0.75% for Adsee 799 and Corexit 7664, and ~0.1% for the other four surfactants). This corresponds to the surfactant dose required to initiate micellization in soil-aqueous systems, i.e., the modified cmc. Because surfactants are likely to sorb/precipitate onto soil, the cmc measured in soil-aqueous systems is higher than the cmc in aqueous systems. Cmc values are obtained in independent laboratory experiments, for example, by monitoring the surface tension at varying surfactant doses. At surfactant concentrations greater than cmc, phenanthrene solubilization appears to increase linearly with surfactant dose. These regions of the solubilization curve may be represented by the enhanced solubility expression presented in equation 1. The flattening out of solubilization curves at higher surfactant doses is an experimental artifact resulting from the finite pool of phenanthrene employed in the solubilization experiments. In the presence of adequate HOC, the solubilization curve would have continued to increase linearly with increasing surfactant dose.

For the extension of laboratory results to field soil washing/flushing applications, the apparent cmc (i.e., cmc in the presence of soil) and the extent of solubilization (e.g., S_w^* value) achieved is of paramount interest in selecting suitable surfactants and surfactant doses. If values for cmc, K_{mn} and K_{mc} are available, the extent solubilized may be estimated from equation 1 at any particular surfactant dose. These parameters may be determined from independent experiments or may be reported in the literature for a particular soil-water-surfactant-HOC system. Subsurface contamination generally involves finite HOC concentrations in soil, so that the type of solubilization curve obtained from laboratory-scale batch experiments employing fixed HOC amounts (Figure 1) is also expected from field studies. In other words, it is likely that similar saturation effects will be observed in field applications. Knowledge of the total soil-sorbed HOC concentration is required to estimate the saturation point for field applications.

Surfactant addition results in a transfer of HOC from soil-sorbed to aqueous micellar phase. Surfactant concentrations used in soil washing/flushing applications need to exceed cmc, but because surfactants are expensive and because they may produce undesirable toxic effects at higher doses, it is desirable to optimize the surfactant dose used.

3.1 *Mathematical programming formulation of the estimation problem*

The relationship between apparent solubility and surfactant dose can be idealized as a piecewise linear curve below a saturation surfactant dose - flat below cmc and then rising with a constant slope. Above the saturation level it asymptotically approaches a maximum apparent solubility.
This can be represented as the function:

$$S(X, S_0, c, a, k) = \begin{cases} S_0 & \text{for } X \le c \\ S_0 + k(X - c) & \text{for } c < X \le a \\ S_0 + k(a - c) + \log X & \text{for } X > a \end{cases} \quad (5)$$

where S is the apparent solubility at surfactant dose X, S_0 is the aqueous solubility below the cmc c, k is the constant slope of the curve above cmc until it reaches a saturation surfactant dose of a, above which the apparent solubility is a logarithmic function of the surfactant dose. Figure 2 presents a schematic representation of the function.

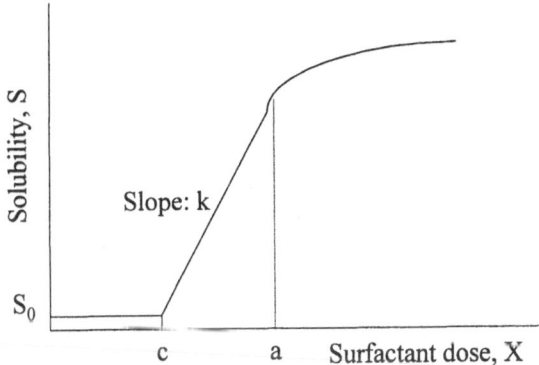

Figure 2. Solubility versus Surfactant dose function

Given a set of observations $\{(S(X_i), X_i)\}$ the objective is to estimate the parameters of interest: S_0, c, k, and a. For a particular estimate of the parameters the predicted value $\hat{S}(X_i)$ may be computed. The minimum square esimator (MSE) is computed as the choice of the parameters that minimizes the sum of the squared error:

$$SSE = \sum [\hat{S}(X_i) - S(X_i)]^2 \quad (6)$$

The objective function (*SSE*) is non-linear. Solubility is a piecewise non-linear function of surfactant dose and appears as constraints in the optimization problem. These constraints may be represented using indicator functions defined over boolean variables. Hence the formulation of the parameter estimation problem is a non-linear mixed integer programming problem. However, the number of observations available from laboratory batch experiments or field studies are typically limited (less than 100). Exploiting this fact, the problem can be solved in a reasonable time using the Generalized Reduced Gradient nonlinear optimization code developed by Leon Lasdon, University of Texas at Austin, and Allan Waren, Cleveland State University. Microsoft Excel solver uses this approach for non-linear optimization and was used in this study. The wide availability of Excel makes this a particularly attractive choice.

In order to demonstrate the viability of this approach, the following data set of 30 observations was used:

Surfactant Dose (mg/l)	Apparent HOC Solubility mg/l
0	0.36
10	1.12
20	1.64
30	1.60
40	1.87
50	-0.09
60	50.88
70	103.96
80	143.64
90	196.03
100	278.19
110	321.71
120	418.33
130	405.45
140	435.12
150	509.04
160	561.84
170	587.50
180	631.27
190	693.51
200	733.52
210	754.40
220	753.12
230	753.53
240	753.18
250	752.21
260	752.04
270	751.10
280	752.79
290	753.34
300	753.68

Based on this data set the MSE estimates of the parameters were $S_0=1.34$ mg/l, CMC $c=48$ mg/L, Slope $k=4.88$, and saturation surfactant dose $a=201$ mg/L. The minimum sum of the squared error attained under these parameter values was 6529. Independent experimental measurements of these parameters yielded $S_0=1.3$ mg/l, $c=55$ mg/l, $k=4.5$, and saturation surfactant dose $a=190$ mg/l. Comparison of MSE estimates with experimentally determined values with 7 other data sets indicated that the discrepancies were typically less than 10%. For all practical purposes this degree of accuracy is deemed quite acceptable. Moreover, while the independent laboratory determination typically takes many hours of work, MSE estimates could be obtained in a fraction of a second.

3.2 Prediction based on artificial neural networks

The MSE estimation requires assumptions about the functional form for the relationship between solubility and surfactant dose. While this may be well justified in the linear region between cmc

158

and saturation dose based on the theory presented in section 2, the behavior of the function below cmc and above the saturation point are not as well understood. Since the prediction problem is essentially one of pattern recognition based on available data sets, artificial neural networks may be used for the purpose.

Windows Neural Networks (WinNN) was used for the analysis., It is a windows-based feed forward neural network simulator with back-propagation learning. WinNN trains in batch mode with a variable epoch length. The 30 available observations were divided into a training set of 20 observations and a test set of 10 observations as shown below:

Training Set

Surfactant Dose	Apparent HOC Solubility
(ppt)	(ppt)
0.00	0.0004
0.01	0.0011
0.03	0.0016
0.04	0.0019
0.06	0.0509
0.07	0.1040
0.10	0.2782
0.11	0.3217
0.13	0.4054
0.14	0.4351
0.16	0.5618
0.17	0.5875
0.19	0.6935
0.20	0.7335
0.22	0.7531
0.23	0.7535
0.25	0.7522
0.26	0.7520
0.28	0.7528
0.29	0.7533

Test Set

Surfactant Dose	Apparent HOC Solubility
(ppt)	(ppt)
0.02	0.0016
0.05	0.0001
0.08	0.1436
0.09	0.1960
0.12	0.4183
0.15	0.5090
0.18	0.6313
0.21	0.7544
0.24	0.7532
0.27	0.7511
0.30	0.7537

HOC solubility and surfactant doses were expressed in parts per thousand (ppt) in order to normalize the data and maintain the observations in the range 0 to 1. A single hidden layer with 2 nodes was used. Adding more nodes to the network did not improve performance significantly. The sigmoid transfer function provided the best results. The target error level of 5% was reached within 5000 iteration in a matter of seconds. Figure 3 presents the plot of the neural network prediction and the target data for the test set. Notice that while the predictions are reasonably accurate in the region below cmc, the neural network fails to track the data adequately in the saturation region.

Neural networks were used for predictions with the 7 other available data sets. The root mean square (RMS) errors obtained using the MSE were compared with those obtained for the neural network based predictions. The table below presents these results:

159

RMS Error

Data Set	1	2	3	4	5	6	7	8
MSE Prediction	14.75	18.2	16.34	22.14	10.51	8.68	15.27	12.1
Neural Network	26.48	30.29	28.46	47.32	18.68	20.04	26.32	28.49

Notice that the MSE estimator performs significantly better than the neural networks.

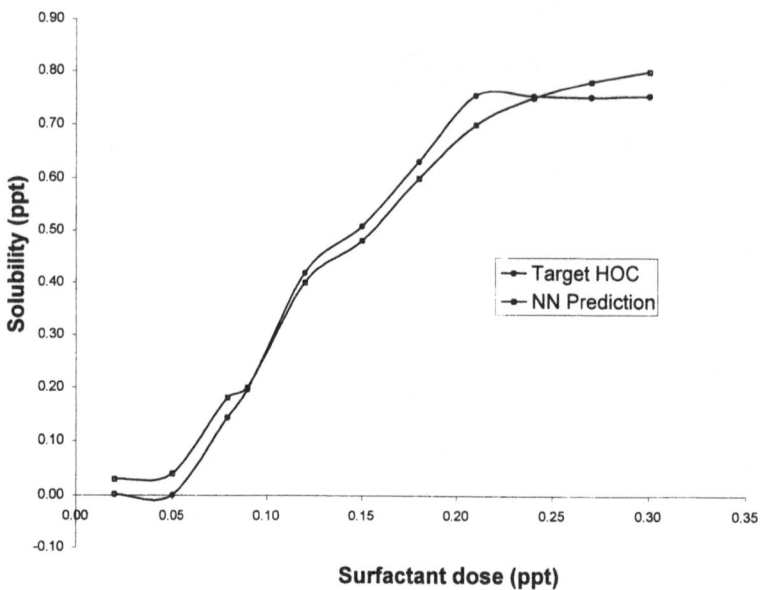

Figure 3. Predictions with Neural Network

4 DISCUSSION

Enhanced pump-and-treat methods have received considerable attention in recent times as more traditional methods for groundwater remediation prove ineffective for target contaminants. The addition of aqueous surfactant solutions to the subsurface have produced increased solubilization of hydrophobic compounds and improved efficiencies of soil washing/flushing applications. In order to predict the performance effectiveness of surfactant-enhanced remediation it is necessary to obtain information on compound hydrophobicity, surfactant properties, and aquifer characteristics. These data are derived from field measurements and laboratory treatability experiments. This paper examines the use of nonlinear mixed integer programming techniques and neural networks at predicting the surfactant solubilization of hydrophobic contaminants.

A conceptual model of HOC partitioning was used to propose a functional relationship between aqueous solubility and surfactant dose. The parameters of this function were estimated using available data so as to minimize the sum of the squared errors. The MSE estimation involves a non-linear mixed integer-programming problem. A Generalized Reduced Gradient nonlinear optimization algorithm implemented in Excel was used to obtain the solutions. The solutions obtained were verified using independent experimental measurements and found to be quite accurate. Hence our proposed MSE estimation technique may be used to obtain parameters of interest rather than using expensive and time consuming experimental techniques.

In order to explore alternate methods for predictions, artificial neural networks were used. While the neural network based predictions are reliable in the constant slope region, the network fails to track the data below cmc and above saturation level faithfully. The RMS errors obtained were significantly higher with neural networks than those obtained with our proposed MSE technique. Further, unlike the MSE technique, the neural network approach does not provide estimates for parameters of interest such as cmc.

While the MSE technique provides an efficient estimation method, appropriate initial parameter 'values must be specified for convergence with this approach. There is a finite probability that the algorithm will get stuck in a local minimum. In most cases reasonable estimates may be provided for initial parameters. However, there may be complex systems where this is not possible; under the circumstances neural network based predictions may prove a viable alternative.

REFERENCES

Chiou, C.T., S.E. McGroddy & D.E. Kile 1998. Partition characteristics of polycyclic aromatic hydrocarbons on soils and sediments. *Environ. Sci. Technol.* 32: 264-269.

Dulfer, W.J., M.W.K. Bakker & H.A.J. Govers 1995. Micellar solubility and micelle/water partitioning of polychlorinated biphenyls in solutions of sodium dodecyl sulfate. *Environ. Sci. Technol.* 29: 985-992.

Hayworth, J.S. & D.R. Burris 1997. Nonionic surfactant-enhanced solubilization and recovery of organic contaminants from within cationic surfactant-enhanced sorbent zones. 1. Experiments. *Environ. Sci. Technol.* 31: 1277-1283.

Hayworth, J.S. & D.R. Burris 1997. Nonionic surfactant-enhanced solubilization and recovery of organic contaminants from within cationic surfactant-enhanced sorbent zones. 2. Numerical simulations. *Environ. Sci. Technol.* 31: 1284-1289.

Kanga, S.A., J.S. Bonner, C.A. Page, M.A. Mills & R.L. Autenrieth 1997. Solubilization of naphthalene and methyl-substituted naphthalenes from crude oil using biosurfactants. *Environ. Sci. Technol.* 31: 556-561.

Karickhoff, S.W., D.S. Brown & T.A. Scott 1979. Sorption of hydrophobic pollutants on natural sediments. *Water Res.* 13: 241-248.

Laha, S. & D.A. Bramwell 1998. Micelle-water partitioning of phenanthrene in aqueous surfactant solutions. To appear in the *Proc. of the Estonian Academy of Sciences.*

Laha, S., Z. Liu, D.A. Edwards, & R.G. Luthy 1995. Surfactant solubilization of phenanthrene in soil-aqueous systems and its effects on biomineralization. In C.P. Huang, C.R. O'Melia & J.J. Morgan (eds), *Aquatic Chemistry: Interfacial and Interspecies Processes,* Advances in Chemistry Series 244. Washington, DC: ACS.

Laha, S. 1992. Solubilization and biodegradation of polycyclic aromatic hydrocarbon compounds in soil-water suspensions with surfactants. Ph.D. Thesis, Carnegie Mellon University, Pittsburgh.

Laha, S. and R.G. Luthy 1992. Effects of nonionic surfactants on the solubilization and mineralization of phenanthrene in soil-water systems. *Biotechnol. Bioeng.* 40:1367-1380.

Laha, S. and R.G. Luthy 1991. Inhibition of phenanthrene mineralization by nonionic surfactants in soil-water systems. *Environ. Sci. Technol.* 25: 1920-1930.

Luthy, R.G., G.R. Aiken, M.L. Brusseau, S.D. Cunningham, P.M. Gschwend, J.J. Pignatello, M. Reinhard, S.J. Traina, W.J. Weber, Jr., & J.C. Westall 1997. Sequestration of hydrophobic organic contaminants by geosorbents. *Environ. Sci. Technol.* 31:3341-3347.

Paya-Perez, A.B., M.S. Rahman, H. Skejo-Andresen & B.R. Larsen 1996. Surfactant solubilization of hydrophobic compounds in soil and water. *Environ. Sci. Poll. Res.* 3(4): 183-188.

Pennell K.D., A.D. Adinolfi, L.M. Abriola, & M.S. Diallo 1997. Solubilization of dodecane, tetrachlorethylene, and 1,2-dichlorobenzene in micellar solutions of ethoxylated nonionic surfactants. *Environ. Sci. Technol.* 31: 1382-1389.

Roy, D., R.R. Kommalapati, S.S. Mandava, K.T. Valsaraj & W.D. Constant 1997. Soil washing potential of a natural surfactant. *Environ. Sci. Technol.* 31: 670-675.

Sun, S., W.P. Inskeep & S.A. Boyd 1995. Sorption of nonionic organic compounds in soil-water systems containing a micelle-forming surfactant. *Environ. Sci. Technol.* 29: 903-913.

Artificial Intelligence and Mathematical Methods in Pavement and Geomechanical Systems, Attoh-Okine (ed.)
© *1998 Balkema, Rotterdam, ISBN 90 5809 028 0*

Modeling of building performance with self learning systems for estimating wind damage

B. Tansel
Drinking Water Research Center and Civil and Environmental Engineering Department, Florida International University, Miami, Fla., USA

I. N. Tansel
Mechanical Engineering Department, Florida International University, Miami, Fla., USA

ABSTRACT: The structural damage to the residential houses in the areas affected by hurricane Andrew was analyzed by two trainable networks, namely Backpropagation Neural Network (BNN) and Abductory Induction Mechanism (AIM). The data used in the study were collected for the hurricane Andrew Damage Assessment program by staff from the Metro-Dade County's Department of Building and Zoning. The input to the neural network programs was divided into an input set consisting of certain characteristics of the house and an output set consisting of the nature of the performance of the house immediately after the hurricane. The input variables included wind speed affecting the house, the type of roof on the structure, the year the house was built, and the zone in which it was located. The output variables were the category of the damage, the percentage of the rood missing, the percentage of the roof trusses missing, and the percentage interior and exterior damage. In all cases, for both neural networks, the average absolute error damage estimates were less than 30 percent for both the training and test cases. The inhabitability, which is a comprehensive damage parameter, was estimated with about 10% error.

1 INTRODUCTION

Hurricane Andrew was the first major hurricane to pass over Dade County since 1926. Maximum sustained surface wind speed (average at 10 meters elevation for 1 minute duration) during landfall over Florida has been estimated at about 145 miles per hour with gusts at 175 miles per hour. These sustained wind speeds correspond to a category 4 hurricane on the Saffir-Simpson scale (Rappaport 1992). It traveled across Biscayne Bay, across developed and agricultural acreage of South Dade County, traversed the Everglades, and exited the peninsula over the islands of the undeveloped southwestern coast, all in about four hours. The hurricane has caused significant damages to roof coverings (tiles and shingles), mobile homes, wood framed walls and roof structures, large metal buildings, boats in marinas and trees. After the hurricane, approximately 135,000 residences needed either roof repairs or a complete overhaul. About 28,000 residences were declared uninhabitable (U.S. Department of Housing and Urban Development 1993, Wakimoto and Black 1991).

Recently, artificial neural networks (NNs) have been successfully used in many different fields (DARPA 1988, Chryssolouris and Giullot 1988). The main advantages of the NNs are their trainability and massively parallel structure. A important potential application for NNs is for cases where future case scenarios could be evaluated from data collected from actual events. With use of NNs, possible damages from future disasters such as hurricanes can be predicted.

In this study, the structural damage data compiled after Hurricane Andrew were analyzed by two neural networks, namely Backpropagation and Abductory Induction Mechanism (AIM). Estimation accuracy of the two trainable networks were compared for a four input and five output analysis. The input parameters were selected based on the construction characteristics of the buildings and wind speed. The outputs were selected as the different types of damages observed at the buildings such as percent of roof covering lost, percent of roof trusses missing, exterior damage, interior damage, and inhabitability of the buildings.

2 BACKPROPAGATION NEURAL NETWORKS (BNN)

Neural network models usually consist of a series of neurons and their respective connections. In every neuron-connection microcosm, the neuron contains an activation and the connection contains the weight. These parameters are in the form of a positive real number and have a maximum value. The NNs system made up of such nodes is inherently dynamic. The processing of data from one state to another is determined by mathematical rules which limit the activation of nodes on the basis of the previous activation of those nodes connected to it. The learning process on NNs takes place along procedures called back propagation. This process consists of two passes: 1. A forward pass where inputs are processed through the network and develop a certain output, and 1. A backward pass where an error signal is produced on the basis of the difference between the actual and expected outputs. As this error is rerun through the NN, the software package strives to gradually reduce it to its minimum point and thereby producing more adequate results.

During the training, NNs can study from only a few cases to millions of cases and establish a model (supervised, unsupervised or self-supervised). The parallel structure of the NN allows use of several processing elements to significantly reduce the computational time in pattern recognition, decision making or simulation studies. Also, the NN can represent the non-linear structures better than the conventional time series models.

The first studies on the development of neural nets started in the 1940s (McCulloch and Pitts 1943, Cowan and Sharp 1988). However, the neural nets gained their present form with the contributions of many researchers from different fields (Amari et al. 1977, Hopfield 1982, Grossberg 1987, Rumelhart et al. 1986).

In NNs, the inputs are connected to the input layer and each connection between the layers has a weight w_{ji} and each node has a logistic activation function. The j th element of the actual output pattern produced by the presentation of input pattern p (O_{pi}) is represented with the following equation (Rumelhart et al. 1986):

$$O_{pj} = \frac{1}{1 + \exp[-(\Sigma w_{ji} O_{pi} + \Theta_j)]}$$

where Θ_j is a bias which works as a threshold. The values of the weights and thresholds can be selected by using the back propagation method given by Rumelhart et al. (Rumelhart et al. 1986). This process consists of two phases. In the first phase the output value O_{pj} is calculated for each unit after the input is presented and propagated forward. The output O_{pj} values are compared with the expected output and the δ_{pj} error signal is calculated. In the second phase, the error signal is passed through the network backward and weight changes are made.

The user has to decide the number of hidden nodes and number of units in each node when the back propagation method is used. This selection has to be made very carefully. The system cannot model the given information if it has too few hidden units; however, too many hidden units does not force the program to generalize the rules and the model would not work well when a new data set is presented.

3 ABDUCTORY INDUCTION MECHANISM (AIM)

The Abductory Induction Mechanism (AIM) (Drake 1991, AbTech 1990) type network was used in this study. AIM is a highly automated non-parametric modeling approach that integrates regression and networking concepts. Backpropagation-type (Rumelhart *et al.* 1986) neural networks have been used in most applications. Training of Backpropagation-type neural networks requires a long time and it is necessary to test various hidden layers with different sizes until the best network is obtained. In this study, the feasibility of AIM (Drake 1991, AbTech 1990, Drake *et al.* 1991) was tested on the encoded data. The advantage of using AIM is that it can test many different possible models automatically until it finds the best one to create an analog output that would represent the wear level. On the other hand, it is similar to supervised neural networks and it requires extensive training to work effectively.

AIM is an advanced learning tool that automatically discovers solutions to complex decision, prediction, control and classification problems. AIM uses abductive reasoning from general principles to specifics under uncertainty. AIM uses small modules to describe the relationship between input and output variables. Between the inputs and each output, there are several modules that are distributed to several layers. Each module has one output and they are called single, double, or triple according to how many inputs it has. Single, double, and triple have one, two, and three inputs, respectively. At the lowest layer, each module is connected to either one, two, or three inputs. The outputs of these modules are connected to the inputs of the modules of the following layer. Some of the modules of a layer may be missing. This architecture continues until the output of the last module in the highest layer is connected to the output of the system. The most complicated module with three inputs and one output has the following format:

$$
\begin{aligned}
\text{Output} = w_0 &+ (w_1\, x_1) + (w_2\, x_2) + (w_3\, x_3) + (w_4\, x_1^2) + (w_5\, x_2^2) \\
&+ (w_6\, x_3^2) + (w_7\, x_1^3) + (w_8\, x_2^3) + (w_9\, x_3^3) + (w_{10} x_1 x_2) \\
&+ (w_{11}\, x_1\, x_3) + (w_{12}\, x_2\, x_3) + (w_{13}\, x_1\, x_2\, x_3)
\end{aligned}
\tag{2}
$$

where w_0, w_1, ..., w_{13} are the coefficients of the module that are estimated by AIM. The AIM program evaluates all the possible combinations by using different modules and finding the least square of error for each case. The final target is to establish a network of modules between the inputs and outputs that has the lowest sum of squares and the least complex structure. Since these two goals conflict, the best combination is selected by considering a user-selected complexity penalty. The AbTech Corporation's AIM program was used in this study.

4 DATA PROCESSING AND ENCODING

The damage data used in the analyses were compiled by Metro-Dade County Building and Zoning Department during the post hurricane damage assessment inspections. After the hurricane, the assessment teams investigated primary structural systems of buildings such as systems that support the building against all lateral and vertical loads experienced during hurricane. In September 1992, Federal Insurance Administration, at the request of the FEMA Disaster Field Office staff, in Miami, assembled a Building Performance assessment Team. Observations were made of damaged and undamaged buildings of similar construction to determine failure conditions. Important observations were made about exterior architectural systems, e.g., roofing components, windows, and doors (FEMA/FIA 1992). In this study four types of structural damages were analyzed: 1. Percent roof missing, 2. Percent roof truss missing, 3. Percent exterior damage, 4. Percent interior damage, and inhabitability.

Roof damages were compiled as percent of roof missing and percent of roof trusses missing. The percent roof missing was reported in relation to the amount of roof cladding missing which includes the underlayment materials such as plywood sheeting, felt and the top most coverings such as tiles and shingles. The damages to the roof trusses were compiled based on the damages to the overall roof framing such as truss bridging and system-wide bracing.

Exterior damage assessments were compiled based on damaged exterior load bearing walls (i.e., walls that support roof framing), and non-load bearing wall panels, damages to the exterior doors and windows. Interior damage assessments were reported based on the extent of damage to the interior walls and doors.

Inhabitability of the buildings were reported based on the overall integrity of the structures. The integrity of the building depends not only on the strength of the components, but also on the adequacy of the connections between them. The inhabitability of each house was distinguished in three categories as uninhabitable and unrepairable, uninhabitable and repairable, and inhabitable.

5 RESULTS

Two different self learning systems were used to model the pattern of damage caused by Hurricane Andrew in South Dade County, Florida: Backpropagation type neural network and Abductory Induction Mechanism neural network. Both programs were trained to "predict" or "assume" the damage conditions to a unit. The data points corresponded to a residence with specific characteristics. For each residence nine parameters were specified, four parameters as input and five parameters as output. In Backpropagation neural network analysis, there was one hidden layer with six hidden nodes. Table 1 presents the input and output parameters and Table 2 presents the characteristics of these parameters.

The available data base consisted of hundreds of cases in 14 different zones. For this analysis, five cases were selected from each zone and entered to the program for training until the errors leveled off. After the training of the NN, the adequacy of the training was tested for twelve new cases from the damage assessment data. Tables 3 and 4 present neural network and AIM output statistics for the training and test data, respectively. Figure 1 presents the comparison of the training and test cases for the two NNs for roof damage.

The parameters predicted by the NNs were generally close to their reported field values, but occasionally showed a significant discrepancy. This can be explained partly by the normal error inherent in the NNs. This is the error point at which the network levels off during the learning process. However, an additional source of error in this case is the classification process of the houses themselves. Since the data used was compiled by many individuals, the classifications given to the residences can vary according to their individual judgments.

In all cases, for both NNs, the average absolute error damage estimates were less than 30 percent for both the training and test cases. The inhabitability, which is a comprehensive damage parameter, was estimated with about 10% error. Considering the subjective nature of the data collection process, the estimates were adequate for this study.

The BNN resulted into lower average absolute errors for inhabitability, percent roof missing, and percent interior damage for the training cases. The AIM resulted into lower

Table 1. Input and output parameters for the neural network.

Input Parameters	Output Parameters
- roof type - location - year of construction - wind speed	- inhabitability - percent of roof missing - percent of roof trusses missing - percent interior damage - percent exterior damage

Table 2. Characteristics of the neural network input parameters.

Parameter	Range	Characteristics
Wind speed	100-200 miles/hour	100-200 miles per hour
Roof type	1-3	1=tile, 2=shingle, 3=gravel
Zone	12-45	The area was divided into 45 zones based on zip codes and specific construction characteristics.
Construction year	1900-1992	
Inhabitability	0-2	0=uninhabitable, unrepairable 1=uninhabitable, repairable 2=inhabitable
Roof missing	0-100 percent	
Roof truss missing	0-100 percent	
Interior damage	0-100 percent	
Exterior damage	0-100 percent	

Table 3. Output parameter errors of the training cases used for BNN and AIM analyses

Neural Network Statistics	Inhabitability	Percent roof missing	Percent roof truss missing	Percent exterior damage	Percent interior damage
Maximum absolute error	0.88	52.00	61.28	45.68	273.63
Average absolute error	0.11	10.60	18.40	15.83	13.24
Standard deviation	0.15	9.90	12.86	10.87	16.12

AIM Statistics	Inhabitability	Percent roof missing	Percent roof truss missing	Percent exterior damage	Percent interior damage
Maximum absolute error	1.000	48.265	60.144	26.067	42.528
Average absolute error	0.167	17.504	15.511	11.657	19.202
Standard deviation	0.009	19.148	18.942	12.942	22.405

average absolute errors percent roof truss missing, and percent exterior damage. With both NNs, the average absolute errors for the training cases were less than 20%. For the test cases, AIM was slightly better in terms of average absolute errors, in comparison to BNN.

Table 4. Output parameter errors of the test cases used for BNN and AIM analyses.

Neural Network Statistics	Inhabitability	Percent roof missing	Percent roof truss missing	Percent exterior damage	Percent interior damage
Maximum absolute error	0.88	34.00	35.90	29.56	65.71
Average absolute error	0.10	12.80	20.87	30.43	23.39
Standard deviation	0.14	10.00	8.36	14.78	30.20

AIM Statistics	Inhabitability	Percent roof missing	Percent roof truss missing	Percent exterior damage	Percent interior damage
Maximum absolute error	1.000	48.265	60.144	26.067	42.528
Average absolute error	0.167	17.504	15.511	11.657	19.202
Standard deviation	0.009	19.148	18.942	12.942	22.405

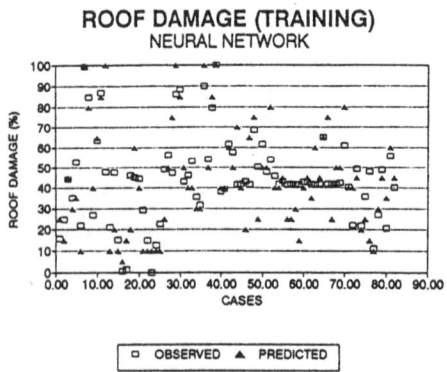

Figure 1.a. Training data for percent roof damage (BNN).

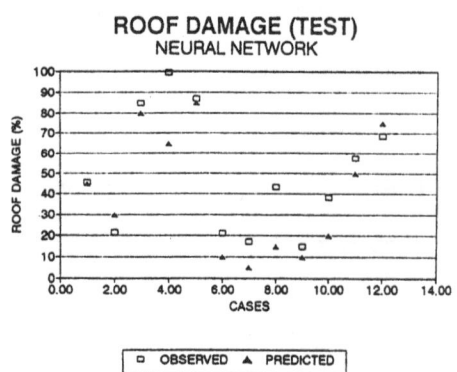

Figure 1.c. Test data for percent roof damage (BNN).

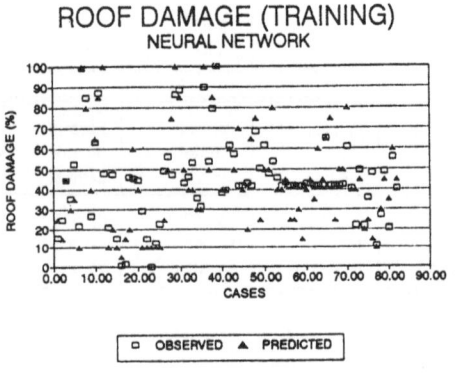

Figure 1.b. Training data for percent roof damage (AIM).

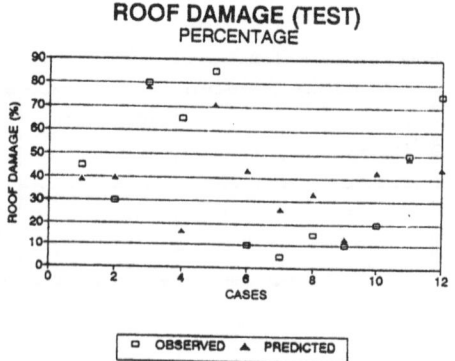

Figure 1.d. Training data for percent roof damage (AIM).

Figure 1. Comparison of training and test cases with BNN and AIM for percent roof damage.

REFERENCES

AbTech Corporation. 1990. *AIM User's Manual*, Charlottesville, Virginia.

Amari, S.K., K. Yoshida & K. Kanatani 1977. A mathematical foundation for statistical neurodynamics", SIAM J. Appl. Mechanics. 33(1):95-126.

Chryssolouris, G. & M. Guillot 1988. An A.I. approach to the selection of process parameters in intelligent machining", *Sensors and Controls for Manufacturing*. Ed. E. Kannatey-Asibu, Jr., Y. Koren, J.L. Stein, Pub. by ASME, New York, PP.199-206.

Cowan, J.D. & D.H. Sharp 1988. Neural nets and artificial intelligence. *The Artificial Intelligence Debate*. Edited by S.R. Graubard.

DARPA, 1988. *Neural Network Study*. AFCEA International Press, Fairfax, Virginia.

Das, M.K. & S.A. Tobias 1967. The relation between the static and the dynamic cutting of metals. *Int. J. MTDR*. 7:63-69.

Drake, K.C. 1991. Highly-automated, non-parametric statistical learning for autonomous target recognition. *Proc. SPIE 20th Applied Imagery Pattern Recognition Workshop*, pp.1-10.

Drake, K.C., T.E. Petty & G.J. Montgomery 1991. *Aeronautical systems applications of advanced technology (ASAAT): Final Report*. ASD/XRF, Wright-Patterson AFB, Contract Number F33657-90-C-2232.

FEMA/FIA (Federal Emergency Management Agency/Federal Insurance Administration) 1992. *Building performance: Hurricane Andrew in Florida; Observation, Recommendations and Technical Guide*. Miami, Florida.

Grossberg, S. 1987. *The Adaptive Brain, I&II*. Amsterdam, Holland.

Hopfield, J.J. 1982. Neural networks and physical systems with emergent collective computational abilities. *Proc. National Academy of Sciences*. USA. Vol. 79, pp. 2554-2558.

McCulloch, W.S. & W.H. Pitts 1943. A logical calculus of the ideas immanent in nervous activity. *Bull. Mathematical Biophysics*. 5:115-121.

Rappaport, E. 1992. *Preliminary Report: Hurricane Andrew (1992)*. (unpublished) National Hurricane Center, Coral Gables, Florida.

Rumelhart, D.E., G.E., Hinton, & R.J. Williams 1986. Learning internal representations by error propagation. *Parallel Distributed Processing: Explorations in the Microstructure of Cognition*, Vol. 1, D.E. Rumelhart and J.L. McClelland (Eds.), MIT Press, pp. 319-362.

U.S. Department of Housing and Urban Development, Office of Policy Development and Research. 1993. *Assessment of damage to single-family homes caused by hurricanes Andrew and Iniki*. Washington, D.C.

Wakimoto, R.M. & P.G. Black 1991. Damage survey of hurricane Andrew and its relationship to the eyewall. *Bull. Am. Meteorological Soc*. 11(4):189-200.

Artificial Intelligence and Mathematical Methods in Pavement and Geomechanical Systems, Attoh-Okine (ed.)
© *1998 Balkema, Rotterdam, ISBN 90 5809 028 0*

Characterization of microscopic oil-polyectrolyte interactions during coagulation by image processing

V. Dimitric-Clark
Civil and Environmental Engineering Department, Florida International University, Miami, Fla., USA

B. Tansel
Drinking Water Research Center and Civil and Environmental Engineering Department, Florida International University, Miami, Fla., USA

ABSTRACT: Coagulation process destabilizes the colloids and other suspended solids by addition of chemicals (coagulants) to agglomerate the suspended particles into flocs which could be removed by gravity settling. During coagulation of oil-water emulsions, small droplets of oil are attached on flocs. The density and size of oil droplets attaching on flocs in comparison to distribution of oil droplets in water phase can be used as a measure of the effectiveness of a coagulant. In this study microscopic image processing techniques were used to characterize the oil droplet-floc interactions to determine effectiveness of coagulation process for treatment of oil-water emulsions.

INTRODUCTION

Petroleum hydrocarbons (PHCs) generally are hydrophobic liquids which at environmental pressures and temperatures exhibit limited solubility in water (Rubin and Mechrez 1989). Although, PHCs are usually assumed to be immiscible, sometimes significant quantities of PHCs may be introduced into the water phase due to solubility of light fractions as well as emulsification processes (Rubin & Mechrez 1989, Eganhouse & Calder 1976, Laughlin et al. 1975). Solubility of petroleum hydrocarbons in water generally decreases as the carbon number increases for both chain and ring PHCs (Jordan & Payne 1980).

A number of physical and chemical treatment processes have been used in treatment of ground and surface waters contaminated with fuel oils. The most commonly used treatment methods include air stripping, dissolved air flotation, carbon adsorption, chemical oxidation, and gravity oil-water separation. Each one of these technologies can be effective within a certain range of process parameters (i.e., concentration of contaminant, type of contaminant, presence of other water quality parameters e.g., suspended solids).

Coagulation is a well known method for removing turbidity and suspended solids from contaminated waters. Coagulation process destabilizes colloids and other suspended solids by addition of chemicals (coagulants) which agglomerates the suspended particles into flocs. In water suspensions, small particles tend to have a net electrical charge, balanced by ions of opposite charge in solution in the vicinity of the particle surface (Veissman & Hammer 1985). The primary sources of surface charge are ionization of surface groups (such as hydroxyl or carboxyl groups), isomeric substitution in the solid lattice, and preferential adsorption of ions or ionizable species from the suspending medium. The surface charge generates a repulsive force between particles, keeping them apart. For larger particles, gravitational forces are dominant and making the particles unstable with respect to the suspension. However, for particles of colloidal size, electrostatic and microhydrodynamic forces dominate and making the suspension relatively stable (Halverson & Panzer 1985). The purpose of coagulation process is to overcome the electrostatic repulsive forces and cause small particles to agglomerate into larger particles, so that gravitational and inertial forces will predominate and affect the settling characteristics of the particles.

Coagulation process involves the addition of a coagulant to the suspension and rapid mixing to destabilize the colloids and other organic contaminants in the mixture. A short rapid mixing phase is followed by slow mixing to promote agglomeration of the suspended particles to form larger flocs. Primary mechanisms by which coagulation can remove organic contaminants include colloid destabilization, precipitation, and coprecipitation (Appiah 1990, AWWA Coagulation Committee 1989). Colloidal destabilization allows the removal of colloids by forming large flocs which can be easily settled or filtered. Removal of organic contaminants by coagulation depends on the extent to which the dissolved contaminants adsorb physically or electrochemically to other particles (i.e., chemical complexation, electrostatic attraction, or hydrogen bonding). Adsorption of the organics onto flocs is more effective than precipitation, because even the trace quantities of the dissolved organics present in water could be removed (Reynolds 1982, Regula et al. 1992, Regula et al. 1993). Polymers, specifically the organic polymers, destabilize colloids by charge neutralization and interparticle bridging. Organic polymers used in water treatment are comprised of specific monomers linked linearly or in branched configurations. These polymers contain functional groups arranged along the chain, and can be charged (polyelectrolytes) or neutral. Since the competing reactions for adsorption and precipitation do not occur for organic polymers, high mixing intensities are not necessary (AWWA Coagulation Committee 1989). After coagulation, some flocs remain in the solution and are usually separated by filtration. Objectives of this study were to characterize removal of oil droplets from oil-water emulsions by polyelectrolytes, and to measure effectiveness of different coagulants in terms of number and size of oil droplets attaching on flocs. During coagulation of oil-water emulsions, small droplets of oil can be attached on flocs. The density and size of oil droplets attaching on flocs in comparison to distribution of oil droplets in water phase can be used as a measure of the effectiveness of a coagulant. In this study image processing techniques were used to characterize the oil droplet-floc interactions.

1 MATERIALS AND METHODS

Surface water samples obtained from a pond located next to Florida International University, Miami, Florida, were contaminated with regular unleaded gasoline at specific concentrations ranging from 100 ppm to 1,000 ppm. The pond water was characterized by turbidity and pH measurements. Unleaded gasoline was characterized by Gas Chromatography analysis and contained petroleum hydrocarbons ranging from C_7H_8 to $C_{21}H_{44}$. Figure 1 presents a typical Gas Chromatography analysis of unleaded gasoline.

Initial coagulant screening tests were performed with different coagulants to determine the effectiveness of the coagulants and optimal coagulant dosage for the oil-water emulsions used. Coagulant screening was conducted by jar tests to identify the most effective coagulants in terms of floc formation rate, floc size, and turbidity removal. Coagulants were prepared by adding 1 milliliter of polymer to 100 milliliters of distilled water for ease of handling as recommended by the manufacturer. Table 1 presents the characteristics of the coagulants used.

Water samples were prepared by mixing the oil-water mixture at specific oil concentrations for 1 hour prior to coagulation experiments. After the coagulants were added, the mixtures were stirred at 100 rpm for 2 minutes (rapid mixing), and then at 35 rpm for 15 minutes, and another 15 minutes at zero rpm for settling of flocs. Microscope slides were prepared with water drops taken from the coagulated and noncoagulated water for microscopic image analysis.

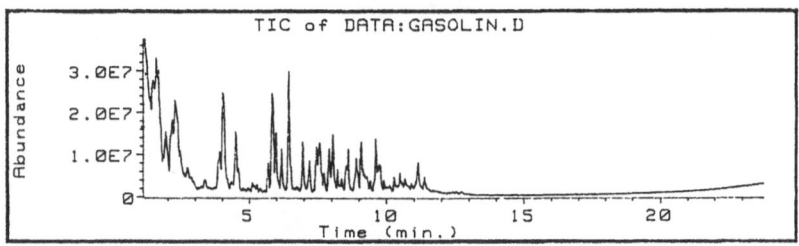

Figure 1. Gas Chromatography analysis of unleaded gasoline.

Table 1. Physical and chemical properties of Cat floc 2953 and EB-5000.

Parameter	Cat floc 2953	EB-5000
Chemical description	Acidic aqueous solution	Acidic aqueous solution
Product class	Cationic coagulant	Cationic coagulant
Ingredients	Poly(dimethyldiallylammonium chloride) ~26% by weight Hydrochloric acid 1-5% by weight Sulfuric acid 1-5% by weight	Poly(dimethyldiallylammonium chloride) ~25% by weight
Flash point	>200°F	>200°F
Boiling point	>212°F	Not available
Solubility in water	Complete	Complete
Specific gravity	1.18-1.20 at 25°C	1.15-1.20 at 25°C
pH	2.5-3.0 at 25°C	3.5-3.8 at 25°C
% volatile by weight	Not determined	65
Appearance	Clear, pale yellow, slightly viscous liquid	Clear to slightly hazy, pale yellow liquid

173

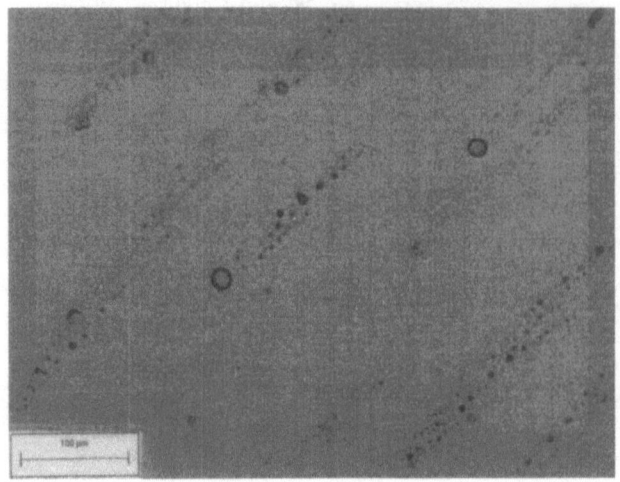

view of oil-water emulsion with no coagulant

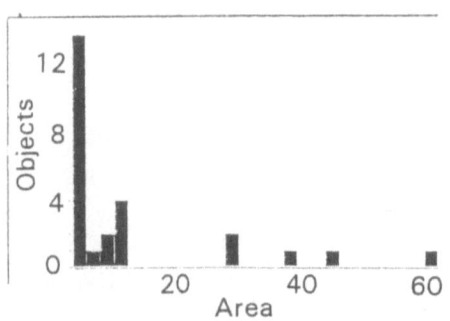 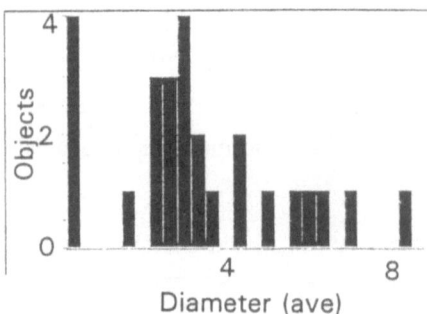

particle characterization histograms

Stats	Area	Diameter
Min	3.838772	0
(Obj.#)	616	273
Max	62.1881	8.827438
(Obj.#)	1861	1861
Range	58.34933	8.827438
Mean	13.6867	3.347291
Std.Dev	15.13497	2.194732
Sum	355.8541	87.02956
Samples	26	26

statistics

Figures 2. Microscopic image of the 500 ppm oil-water emulsion contaminated with 500 ppm unleaded gasoline without coagulant.

2 IMAGE PROCESSING

The microscopic images were captured and analyzed by Image-Pro Plus image analysis software (Image-Pro Plus 1997). The image obtained was digitized into pixels for

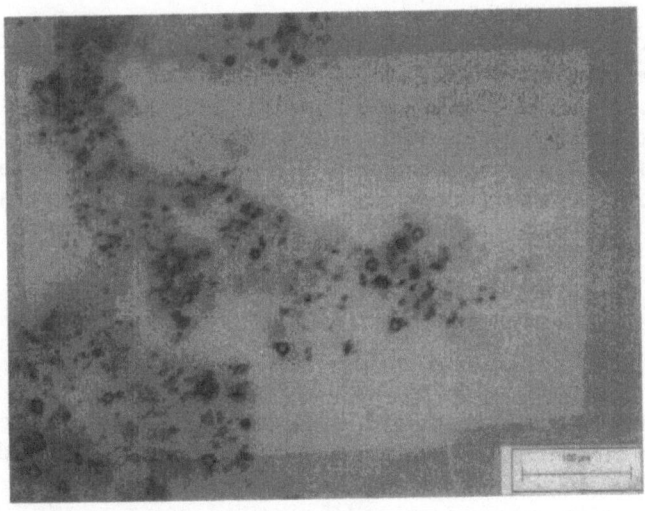

view of oil-water emulsion with coagulant

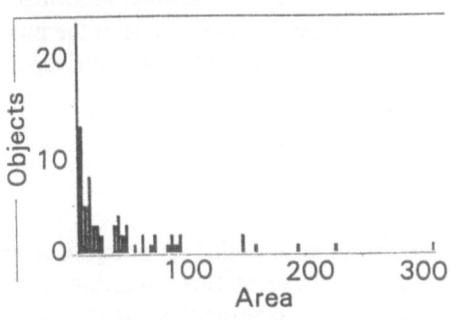 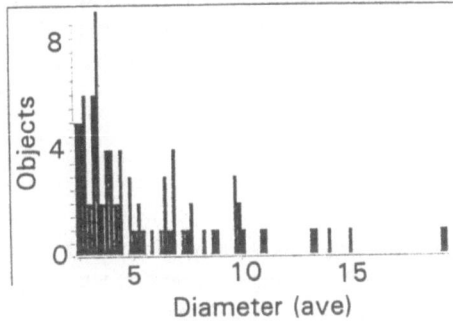

particle characterization histograms

Stats	Area	Diameter
Min	3.838772	2.283697
(Obj.#)	47	47
Max	314.0115	19.91781
(Obj.#)	155	156
Range	310.1727	17.63412
Mean	37.84499	5.726713
Std.Dev	53.34654	3.726947
Sum	3292.514	498.2241
Samples	87	87

statistics

Figures 3. Microscopic image of the 500 ppm oil-water emulsion contaminated with 500 ppm unleaded gasoline with using Cat floc 2953 at the optimum dosage.

analyses. The bitmap of the image can be filtered by convolution (linear) or nonconvolution (nonlinear) filters to adjust the sharpness of the image to distinguish the specific characteristics of the image. The data from the image can be collected by intensity analyses in the form of:

- histogram analysis to create a histogram of the image,
- line profile analysis to pot intensity values along a given line, and
- bitmap analysis to display the values of individual pixels in a bitmapped image.

The image can be measured either automatically or manually. Once the objects are defined, spacial and intensity measurements can be performed. Typical measurements menu includes characteristic measurements of the objects such as area, diameter, length, optical density, radius ratio, roundness, as well as characteristics of the defined view area.

3 CHARACTERIZATION OF OIL-POLYELECTROLYTE INTERACTIONS

Figures 2 and 3 present the microscopic images at 100x magnification of the oil-water emulsion samples contaminated with 500 ppm unleaded gasoline without and with coagulant, respectively. The sample shown in Figure 3 was prepared after jar tests using Cat floc 2953 at the optimum dosage (4 ml of 1% coagulant solution per liter of pond water--i.e., 0.04 ml of pure coagulant/liter of pond water). The coagulation of oil emulsions at 500 ppm concentration with coagulant Cat floc 2953 resulted in formation of large white flocs (2-4 millimeters) visible with naked eye. The analysis of the image at 100 magnification showed that oil droplets were entrapped both on the surface and within the floc.

4 CONCLUSIONS

The image analyses performed by Image-Pro software provided information about specific physical characteristics of the interactions between flocs forming during coagulation process and oil droplets. The number and diameter of droplets attached on floc can be used to determine the effectiveness of coagulants for treatment of oil-water emulsion. The image analysis can effectively be used to characterize the coagulation of oil-water systems by conducting a statistical analysis of the floc sizes, floc densities, oil droplet sizes, and oil droplet densities.

REFERENCES

Appiah, A. 1990. Coagulation: rejuvenation for a classical process. *Water/Engineering and Management*. 10: 25-32.

AWWA Coagulation Committee 1989. Committee Report: Coagulation as an integrated water treatment process. *Jour. American Water Works Association (AWWA)*. 81(10): 72-78.

Eganhouse, R.P. & J.A. Calder 1976. The solubility of medium molecular weight aromatic hydrocarbons and the effects of hydrocarbon co-solutes and salinity. *Geochimica et Cosmochimica Acta*. 40:555-561.

Halverson, F. & H.P. Panzer 1985. Flocculating agent, in *Kirk-Othmar Concise Encyclopedia of Chemical Technology*, John Wiley & Sons.

Image-Pro Plus. 1997. *Image-Pro Version 3 for Windows*. Media Cybernetics, Silver Spring, Maryland.

Jordan, E.R. & J.R. Payne 1980. Fate and weathering of petroleum spills in marine Environment. *Ann Arbor Science Publication*. Ann Arbor, Michigan.

Laughlin, R.B., O. Linden & J.M. Neff 1975. A study on the effects of salinity and temperature on the disappearance of aromatic hydrocarbons from the water soluble fraction of No. 2 fuel oil. *Chemosphere*. 10:741-749.

Regula, J., B. Pascual, B. Tansel & R. Shalewitz 1992. Effect of coagulation prior to ultrafiltration for removal of petroleum hydrocarbons. *Proceedings of the South Florida ASCE Annual Conference*. Paper No. 22. Boca Raton, Florida, October 2-3.

Regula, J., B. Pascual, B. Tansel & R. Shalewitz 1993. Coagulation pretreatment for ultrafiltration of petroleum hydrocarbon contaminated waters. *Proceedings of the 48th Purdue Industrial Waste Conference*. 249-259. West Lafayette, Indiana, May 10-12.

Reynolds, T.D. 1982. Unit operations and processes in environmental engineering *PWS Publishers*. Boston, MA.

Rubin, H. & E. Mechrez 1989. Transport of organic pollutants in a multiphase system. *Springer-Verlag*. 231-250.

Viessman, W. & M.J. Hammer 1985. Water supply and pollution control. *Harper Collins Publishers*. New York, New York.

Artificial Intelligence and Mathematical Methods in Pavement and Geomechanical Systems, Attoh-Okine (ed.)
© *1998 Balkema, Rotterdam, ISBN 90 5809 028 0*

System identification – Tutorial

K. K. Yen
Department of Electrical Engineering, Florida International University, Miami, Fla., USA

ABSTRACT: This paper presents the basic concepts of system identification techniques and their importance in model building, the available algorithms, and possible problems that may be encountered in experiment design in civil engineering systems. Based on mathematical models, which may represent a set of equations, most civil engineering problems can be solved. Through studying these models, the behavior of the system can be understood, and used for system performance prediction and control, as and when it is possible.

1 INTRODUCTION

What is system identification, and why is it an important discipline in engineering fields? System identification is the process of deriving a valid mathematical model from observed data for a given physical system in accordance with some predetermined criterion. For example, time-varying wind load acts on a building causing it to oscillate. Studying these oscillations we can infer important mechanical characteristics of the building. Engineering problems are solved most times based on mathematical models, a set of equations. Through studying its model we can understand the behavior of the system, and whenever possible, use it for system performance prediction and control. In this sense modeling means the study of the mechanisms inside a system. In general, models are not exactly equal to the physical system, and they, no matter how complex, only approximations of the system. For different purposes of study, we will deal with different system variables and make different kind of assumptions and approximations, so different models will be developed. This explains why the model for a physical system is not unique and depends on the system characteristics to be investigated.

Broadly speaking, modeling methodologies can be classified into two categories. The first one, mathematical modeling method, bases on fundamental physical laws and interconnections between components and involves three separate tasks:

(1) identifying and idealizing of individual elements,
(2) identifying and idealizing of their interactions, and
(3) applying basic laws and relations systematically

The measurement of input and output data sets will be utilize for coefficients estimation, then a complete model is inferred. For complex systems, the work involved becomes overwhelming and the resulting model may not have any practical value. The alternate is the experimental modeling method. In this method, we treat the system as a "blackbox", provide it with some stimuli, and measure its response. This modeling process involves selection of a model from a set of popular mathematical relationships, determination of its order, and estimation of the value of model parameters, so the developed model will best fit the observed input-output data.

Various identification methods, employing different concepts concerning the form of the identification model, exist. Each one of them has its own ranges of applicability. Identification methods suitable for implementation on a digital computer are Fourier analysis, correlation analysis, spectral analysis, model fitting, and parameter estimation. The most widely used parameter estimation methods are those based on the principles of least-squares and the maximum-liklihood methods.

Modeling and identification is a very broad subject and is well developed in control engineering field. Its applications have been also extended outside the traditional engineering disciplines such as the studies of social systems, economic systems, or biological systems etc.[1] Today many user-friendly software packages for system identification in which the theories are packaged are available. One of them is System Identification Toolbox used with MATHLAB.[2] This Toolbox contains all the common techniques to adjust parameters in linear models. It also allows you to examine the model's properties to check if they are good, as well as to preprocess the measured data. Also many textbooks and tremendous research papers on this subject are available. Here we can only cite a few. [3-10] Therefore, in this paper the purpose is to introduce the basic concepts and algorithms'to engineers in civil engineering to stimulate the possible applications in their field. To achieve this goal the focus will be moved from algorithms development to the understanding of the possibilities and limitations of identification.

2· IDENTIFICATION PROBLEMS

A typical engineering problem with all signals is shown in Figure 1. The output signal $y(t)$, contaminated by possible disturbance $w(t)$ and noise $v(t)$, conveys information about the system. For example, the oscillation of a bridge due to wind gusts act on it will disclose important mechanical characteristics of the bridge. In general, for different purposes engineering problems can be classified into two categories; they are

    (4) forward problems
          (a) analysis – to predict the performance of a system with known model and different inputs
          (b) design – to modify a system to achieve predefined performance
    (5) inverse problems
          (a) identification – to identify the characteristics of a system from collected data of input and output signals
          (b) deconvolution – to estimate the cause for a known system and its output

Two major tasks has to be performed in the solving system identification problems using canned software packages. They are

    (1) selection of a model structure and determination of one in the structure which satisfactorily fits the measured data sets, and
    (2) transformation of the determined model to a desired form for parameter identification.

A model structure is a family of models with adjustable parameters such as differential equations, difference equations, transfer functions, state-space models etc. In system

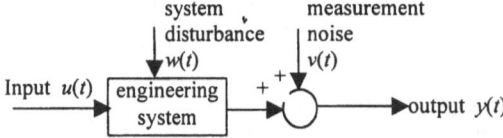

Figure 1 System diagram

identification these parameters will be calculated to minimize the difference between the predicted response and the observed one. So a system identification problem involves finding a good model structure, the order of the model, and good numerical values of its parameters.

The availability of the degree of a priori knowledge, such as the dimension of the system and the structure of the model, regarding the system is the most important factor in identification problem formulation. It has been noted that identification techniques that assume less a priori knowledge are not only less accurate, but also more complex in terms of their mathematical representation and more time-consuming in the computation than those with more prior knowledge.

3   MODEL STRUCTURES[5, 11, 12]

A system is a collection of objects arranged in an ordered form to achieve a designated goal. A model, on the other hand, consists basically of a set of mathematical equations which represents the essential aspects of the system. The ability of the model to predict the observed response is a measure of the fidelity of the model to the system. For a model to be useful, it must not be so complicated that it cannot be understood and thereby be unsuitable for predicting the behavior of the system. On the other hand, it must not be trivial to the extent that predictions of the behavior of the system are grossly inaccurate.

3.1 Static Models

In static situation, identification is solely for the purpose of recognition. For example, a concrete specimen is loaded and its deformation can be measured to determine material properties such as Young's modulus and Poisson ratio. For a single-input, single-output (SISO) system, we can describe a static model by algebraic relation such as

$$y = F(u, \mathbf{a}) \tag{1}$$

where y is the output, $F(\cdot)$ is an algebraic mapping, u is the input, and the vector a consists of a set of coefficients to be estimated. Here the function $F(\cdot)$ can be either linear or nonlinear. For multiple-input, multiple-output systems, the model can have the form

$$y_1 = F_1(u_1, \ldots u_r, a_{11}, \ldots, a_{1n})$$
$$\vdots \tag{2}$$
$$y_m = F_m(u_1, \ldots u_r, a_{m1}, \ldots, a_{mn})$$

Many problems in civil engineering applications can be represented in this way and the identification process is just coefficient estimation.[13]

3.2 Dynamic Models

Based on system characteristics, models of dynamic systems can be classified in many different ways such as lumped- or distributed-parameter, linear or nonlinear, continuous- or discrete-time, time-varying or time-invariant, signal- or multi-input, and deterministic or stochastic etc. [14, 15] In a lumped-parameter model, the dynamic behavior is described in terms of ordinary differential equations; for distributed-parameter model, partial differential equations are used. Strictly speaking, linear time-invariant systems do not exist in practice since all physical systems are nonlinear and varying with time to some extent. However, for a large number of practical cases, if we carefully limit the signal magnitude and the operating time interval on the basis of past experiences, then a linear time-invariant model may be adequate. Incremental or piecewise linear models are quite often used for the approximate of the nonlinear system.   Other reasons to support the use of linear time-invariant models are due to their simplicity and well-developed methodology in solving them. However, in using such models, one must be careful and should have an idea of the limits of their validity. If nonlinearity is the property to be study, we have to deal with nonlinear model.[16] Even in this situation, sometimes using physical insight about the system we can do nonlinear transformation on the model and end up with a linear one in new variables. Recent research results also show that partial differential equations, integro-differential equations can be reduced to systems of ordinary differential equations and boundary-value problems into initial-value problems.[17, 18] So linear ordinary differential equations can be used to model a large number of dynamic systems.

Almost all of physical systems in engineering are of the continuous-time type, but the application of the digital computer for identification makes it desirable to use discrete-time models. Often the determination of the parameters of a discrete-time model is more straightforward. Furthermore, provided that the sampling interval satisfies certain conditions, the determination of the continuous-time model from the discrete-time model is fairly straightforward.[19, 20, 21]

Based on the above argument, in the following discuss on identification we will limit ourselves to the SISO, linear, discrete-time model as discussed in the Toolbox [2] which are in the form of

(1) difference equations
(2) Transfer function models
(3) State-space models

Difference equation models

Several popular classes of Ready-made models for linear discrete-time systems are described below. [2, 5,7]

(1) ARMAX models

The ARMAX model (autoregressive moving average with exogenous input) is an important class of difference equation of the form

$$A(z^{-1})y(k) = z^{-d}B(z^{-1})u(k) + C(z^{-1})v(k) \qquad (3)$$

where d is the time delay and A, B, C, are polynomials

$$A(z^{-1}) = 1 + a_1 z^{-1} + ... + a_{n_a} z^{-n_a}$$

$$B(z^{-1}) = b_0 + b_1 z^{-1} + ... + b_{n_b} z^{-n_b}$$

$$C(z^{-1}) = 1 + c_1 z^{-1} + ... + c_{n_c} z^{-n_c}$$

with the unknown parameters

$$\left(a_1, ... a_{n_a}, b_0, ..., b_{n_b}, c_1, ..., c_{n_c}, \sigma_v^2 \right)$$

Here $v(k)$ is white noise.

(2) ARX model

The ARX model, a special case of ARMAX model, is also called an infinite impulse response (IIR) model or controlled autoregressive mode and is represented as

$$A(z^{-1})y(k) = z^{-d}B(z^{-1})u(k) + v(k) \tag{4}$$

or

$$\hat{y}(k) = -\sum_{i=1}^{n_a} a_i y(k-i) + \sum_{j=1}^{n_b} b_j u(k-j) \tag{5}$$

where $\hat{y}(k) = y(k) - v(k)$ is the estimated output.

(3) FIR models

The FIR model is another special case of the ARMAX model with the form

$$y(k) = B(z^{-1})u(k) + v(k) \tag{6}$$

or

$$\hat{y}(k) = \sum_{j=1}^{n_b} b_j u(k-j) \tag{7}$$

All the models show us how the output is influenced by the input and disturbance at various time instant. The identification problem is then to use measurements of u, v, and y to determine

    (a) coefficients, a_i and b_i
    (b) number of delayed outputs in the description
    (c) number of delayed inputs to use
    (d) time delay, d, in the system

Transfer Function Models

The transfer function model has the form

$$y(k) = H_u(z)u(k) + H_v(z)v(k) \tag{8}$$

There are several algorithmic reasons to factorize transfer functions into numerator and denominator polynomials.[2, 5, 7] Two popular transfer function models are

(1) Output-Error (OE) models

$$y(k) = \frac{B(z^{-1})}{F(z^{-1})}u(k) + v(k) \tag{9}$$

This model contains no assumption on the spectrum of the disturbance sequence $\{v(k)\}$.

(ii) Box-Jenkins models

$$y(k) = \frac{B(z^{-1})}{F(z^{-1})}u(k) + \frac{C(z^{-1})}{D(z^{-1})}v(k) \tag{10}$$

This model contains a noise model with the white-noise sequence $\{v(k)\}$ filter through the transfer function $C(z^{-1})/D(z^{-1})$. One attractive property of the Box-Jenkins model is that it yields separate descriptions of the input-output relationship between u and y and the noise spectrum model described by $C(z^{-1})/D(z^{-1})$ and $\{v(k)\}$.

The model structures for the above Read-made models are given in Fig.2.

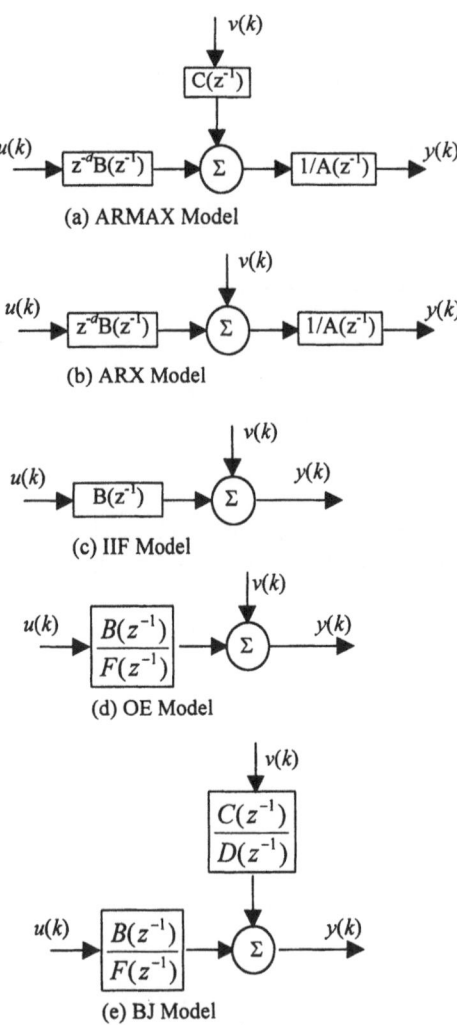

Figure 2: Model Structures

All of the described models can be formulated as linear regression of the form

$$y(k) = \mathbf{w}^T \Phi(k) + v(k) \tag{11}$$

assuming that, $y(k) = \hat{y}(k) + v(k)$. For example, the ARX model can be considered as the form

$$\mathbf{w} = [a_1, ..., a_{n_a}, b_1, ..., b_{n_a}]^T$$

and

$$\Phi(k) = [-y(k-1), ..., -y(k-n_a), u(k-1), ..., u(k-n_b)]^T$$

In many cases of system identification, the effects of the noise on the output are insignificant compared to those of the input. With good signal-to-noise ratios (SNR), it is less important to have an accurate noise model.

State-Space Models[11, 14, 15]

The state-space model can be described by

$$\mathbf{x}(k+1) = \mathbf{\Phi}\mathbf{x}(k) + \mathbf{\Gamma}\mathbf{u}(k) + \mathbf{v}(k)$$
$$\mathbf{y}(k) = \mathbf{C}\mathbf{x}(k) + \mathbf{w}(k)$$
(12)

with $\mathbf{x}$-$n\times1$ state vector, $\mathbf{u}$-$r\times1$ control vector, $\mathbf{y}$-$m\times1$ output vector, $\mathbf{v}$-$n\times1$ disturbance vector, and $\mathbf{w}$-$m\times1$'measurement vector. The conversion of the model from continuous-time to discrete-time is derived in the given reference. The state vector $\mathbf{x}$ is not measured, but can be reconstructed from the measured output data if the system is observable.

4   IDENTIFICATION METHODS

Identification methods can also be classified in many ways such as parametric methods or nonparametric methods, batch methods or recursive methods, on-line methods or off-line methods etc.[3, 4, 22] If we know the model structure in advance, then parametric methods can be utilized; otherwise, nonparametric procedures have to be employed. Algebraic equations, difference equations, and transfer functions are examples of parametric models. Parametric methods estimate parameter values of given models in such a way that the best fit between the model's predicated output and the measured one is achieved. Nonparametric methods on the other hand estimate model behavior based on a history of input-output measurements without necessarily using a model. Typical methods include correlation analysis and spectral analysis. The accuracy from these methods will be improved with an increased number of measurements. The input signals used can be either operating signals of the system or some artificial ones. Regardless of the type of signal employed, measurements must be taken when the system is in its transient state because dynamic parameters cannot be identified when the system is in its steady-state. In general, a parametric model can be obtained from a nonparametric one.[8]

Generally, unknown system parameters can be identified in either batch form or recursive form. The former is mainly for off-line computation and the latter is used for on-line identification. In batch methods we first collect a large amount of input and output and store them in a digital computer, then the estimation process starts to calculate parameters values which will minimize a prescribed cost function. Several advantages are associated with batch methods. They are

(1) choice in the selection of input signals
(2) flexibility in the selection of computational methods
(3) no restriction on computing time
(4) high accuracy of the estimates

The recursive scheme for on-line identification, on the other hand, has the following properties;

(1) it does not require a special input
(2) it requires minimum amount of memory for data storage
(3) it will update the estimates of the parameter after each sampling instant
(5) its computation time is a fraction of the sampling period.

In general, on-line methods will not produce as accurate results as off-line methods. This is because batch methods use a much larger amount of data in their calculation, i.e., more information about the system is utilized. However, in many practical situations one cannot afford to wait for the time required in collecting all the data. As a matter of fact, it will be recognized that life is the art of reaching decisions from insufficient data.

A large variety of methods have been applied to system identification, and they can be classified in many ways; one scheme for classification is given below.[8]

(1) Classical Methods: (mostly off-line)

(a) Frequency Response identification
(b) Impulse response identification by deconvolution
(c) Step response identification
(d) Identification from correlation functions
(2) Equation-error Approaches: (batch-processing)
(a) Least-squares
(b) Generalized least-squares
(c) Maximum likelihood
(d) Minimum variance
(e) Gradient methods
(3) Model Adjustment Techniques:
(a) Least-squares (recursive)
(b) Generalized least squares (recursive)
(c) Instrumental variables
(d) Bootstrap
(e) Maximum likelihood (recursive)
(f) Correlation (recursive)
(g) Stochastic approximation

Algorithms specially designed for nonlinear system identification are also in the literature. [5, 8, 9, 27]. In the following, we will discuss only Kalman filtering method, adaptive filtering method, and neural network method. [23, 24, 25] Other methods are covered in almost every textbook on system identification and will be skipped here.

4.1 Kalman Filtering Method

A typical recursive identification algorithm is

$$\hat{\theta}(k) = \hat{\theta}(k-1) + \mathbf{k}(k)[y(k) - \hat{y}(k)] \tag{13}$$

This algorithm says that the current parameter estimate $\hat{\theta}(k)$ depends on its previous estimate $\hat{\theta}(k-1)$ and a correction which is proportional to the difference between the current measured output $y(k)$ and the estimated output $\hat{y}(k)$ based on previous estimate $\hat{\theta}(k-1)$. The gain vector k(k) determines how the current prediction error [y(k)- $\hat{y}(k)$] will affect the update of the parameter estimate. We assume that the model structures can be written as

$$y(k) = \phi^{T}(k)\theta(k) + e(k) \tag{14}$$

where vector θ contains the unknown parameters and vector ϕ formed by old inputs and outputs. The most logical approach to the adaptation problem is to assume a certain model for how the true parameters θ change. A typical choice is to describe these parameters as a random walk

$$\theta(k) = \theta(k-1) + \mathbf{w}(k) \tag{15}$$

Here the disturbance noise vector $\mathbf{w}(k)$ and the measurement noise $v(k)$ are assumed to be white Gaussian with covariance matrices

$$E[\mathbf{w}(k)\mathbf{w}^{T}(k)] = \mathbf{R}_{1}$$

$$E[e(k)e^{T}(k)] = r_{2}$$

The Kalman filtering method can be used for prediction of θ based on the past data at time k which will minimize the cost function

$$J = E[(\hat{\theta} - \theta)^T (\hat{\theta} - \theta)]$$ (16)

The complete algorithm is

$$\hat{\theta}(k) = \hat{\theta}(k-1) + \mathbf{k}(k)[y(k) - \hat{y}(k)]$$

$$\hat{y}(k) = \phi^T(k)\hat{\theta}(k-1)$$

$$\mathbf{k}(k) = \frac{\mathbf{P}(k-1)\phi(\kappa)}{r_2 + \phi^T(k)\mathbf{P}(k-1)\phi(k)}$$

$$\mathbf{P}(k) = \mathbf{P}(k-1) + \mathbf{R}_1 - \frac{\mathbf{P}(k-1)\phi(k)\phi^T(k)\mathbf{P}(k-1)}{r_2 + \phi^T(k)\mathbf{P}(k-1)\phi(k)}$$

4.2 Adaptive Filtering Method

An adaptive filter can be used in imitating the behavior of a dynamic system. Modeling a SISO system is shown in Figure 3. We drive both the unknown system and

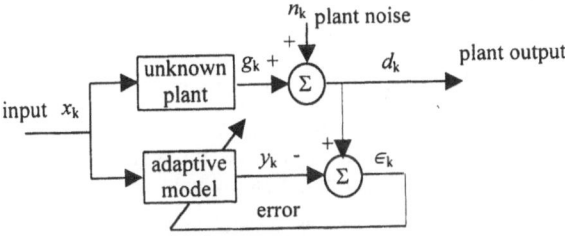

Figure. 3: Modeling SISO Noisy Case

the adaptive filter by the same input. The adaptive filter will adjust itself, so the output will be a best least-squares fit to that of the unknown system. Upon convergence, if the input is wideband and the structure contains enough adjustable parameters, a close fit or perhaps a perfect fit will be possible. In this sense, the adaptive system becomes a model of the unknown system.

In many practical cases, the plant to be modeled is noisy. Internal plant noise appears at the plant output and is commonly represented as an additive noise. This noise is generally uncorrelated with the plant input. It can be shown that the least-squares solution will be unaffected by the presence of the plant noise. This is not to say that the convergence of the adaptive process will be unaffected by plant noise, only that the expected weight vector of the adaptive model after convergence will not be affected. The least-squares solution will be determined primarily by the impulse response of the plant to be modeled. It could also be significantly affected by the statistical or spectral character of the plant input signal.

4.3 Neural Network Method[25]

In practice many systems and processes are nonlinear in nature. Neural networks can be used to model this nonlinearity. For an unknown stable time-invariant causal, discrete-time dynamic system the input and output signals are observable and measurable. We can construct a neural network model which, when subject to the same input $u(k)$, produces an output $y(k)$ which estimates the output $y(k)$ of the system in the sense that the specified cost function

$$E = \frac{1}{2}\sum_k |e(k)|^2 \qquad\qquad (17)$$

of the errors e(k) =y(k) - $\hat{y}(k)$ is minimal. The identification problem can be divided into three successive steps:

    (a) model building,
    (b) the learning procedure (i.e., parameter determination)
    (c) diagnostic checking.

For a SISO nonlinear system the identification models can take the form

$$\hat{y}(k) = F[y(k-1),...,y(k-n_a),u(k-1),...,u(k-n_b)] \qquad (18)$$

where $u(k)$ and $y(k)$ are, respectively, the system input and output, $F(\cdot)$ is the unknown system function and $\hat{y}(k)$ is the estimated vector of $y(k)$. Furthermore, we assume that the system is a continuous functional or mapping of the input, i.e. small changes in the system input result in small changes in the system output. However, this model is rather difficult to learn because of the relatively large number of unknowns involved. Narendra and Parthasarathy [26] has propose some simplified models to avoid this difficulty. All these models can be considered as an extension of the IIR model. For each of the models we can derive an error back-propagation learning algorithm which enable us to find the unknown parameters, i.e., to reconstruct a nonlinear differential equation which fits the unknown dynamic system.

5 DESIGN CONSIDERATIONS

The identification procedure must be carried out under the conditions that are as similar as possible to those under which the model is going to be used. In practice, the main task in the design of an identification experiment can be divided into four major parts; they are

    (1) Data Handling – input data generation, sampling period consideration, data filtering, output data length selection, etc.
    (2) Model Structure Selection – difference equation models, transfer functions, State-space models, etc.
    (3) Algorithm Selection – nonparametric identification methods or parametric estimation methods, etc.
    (4) Model Validation – model verification through simulation, computation and analysis of residuals, comparison between different models' properties, etc.

5.1 Data Handling

The success of an identification experiment heavily depends on whether the collected data contain significant information about the system. Several factors need to be considered; they are

    (1) What kind of signals will be proper inputs?
    (2) Which output signals should be measured?
    (3) What will be the sampling period?
    (4) How many data points need be collected?
    (5) Is data filtering required?

Choice of Input

Since in identification we are interested in certain specific system behavior, appropriate input signals should be selected so that from the resulting response the most accurate model may be identified. Several signals have been widely accepted as test signals in engineering fields; they are impulse, step, sinusoidal, and random noise.[4, 5, 14] A proper input should excite all interesting aspects of the system in order to obtain a realistic model. For example, if we choose the input as a single sinusoid with a fixed frequency ω, then many systems may generate the same output with this input. It is obvious that this is not a good choice. So it is thus important that $u(t)$ contain wide enough frequencies.

Although an ideal step is physically unrealizable, an approximation can be fairly close if the rise-time of the step input is much shorter than the period of the highest frequency of interest in the identification. The levels should be chosen so that they correspond to maximally allowed variations. If the system is nonlinear, an interval for the input that corresponds to a desired operation point should be chosen. When constructing a nonlinear model, it is usually also necessary to work with more than two input levels. If a pulse train with different duration is adopted as the input signal, it is important to keep pulse-width long enough. Otherwise, the response will be hardly visible, that is, just covering a negligible part of the rise time of the step response. Moreover, it should be useful to have occasional pulses that are constant over such long periods that the step response more or less settles. This gives some practical guidelines for the choice of pulse lengths.[7]

For noise inputs, in the absence of any information on them we can assume that $\{v(k)\}$ is white Gaussian noise sequence for simplicity. If this is not true, we may consider it as the output of a linear filter with a unit-variance white Gaussian noise input. Although it is not possible in practice to obtain an ideal white noise input, it can be approximated by noise with power spectral density constant over the frequency range of interest for identification. [5, 27]

Methods for optimal input design have been develop by researchers. One method for linear system is based on the property of Fisher's information matrix. [28] Optimal input can also be design to maximize the sensitivity of the system output to the unknown parameters under the assumption that the input signal is either energy or power constrained. [29, 30] Gagliardi has shown that for a finite set of parameters the input vectors that map into the output set with the largest perimeter in the vector space maximize the probability of identifying the correct parameters. [31] All the derivations of these methods base on the assumption that the period of observation is sufficiently large.

Choice Of Sampling Interval[14, 22]

The selection of sampling interval is determined by the time constants of the system. Data redundancy will happen if we sample the signal considerably faster than the system dynamics. On the other hand, aliasing will be observed if the sampling frequency is too slow. Nyquist Sampling Theorem only provides us with a theoretical lowest bound for bandlimited signals and this information does not have any practical use. A rule of thumb is to choose a sampling frequency about 10 times of the bandwidth of the interest for the modeling. This corresponds approximately to placing about 5-8 sampling points over the rise time of the system's step response. It is thus valuable to first obtain a step response of the system.

If in a situation we are not certain of the choice of sampling frequency, it is wise to sample fast at the data collection. Then the data sequences can be decimated; that is, every rth value of the original data is chosen before they are used in the identification routine. Another popular practice to determine if T is sufficiently small is to let $T = 0.5T$ and repeat the computation. If no appreciable change is noted then T is sufficiently small; otherwise, we let $T = 0.5T$, and repeat. If a prior information is known about the system and its input signals, a low-pass anti-aliasing filter may be considered at the front end to help reduce the sampling frequency.

Data Treatment

It is always a good practice to plot and study the data sets before we process them in order to avoid some so obvious traps. For example, in some cases it is very easy to identify that some measured values are incorrect. If these values were accepted without care, they might create a devastating effect on the estimated models. Segments that contain inaccurate or doubtful measurement values should be avoided. If this is impossible, the inaccurate values could be smoothed by hand to interpolated or predicted values.

It is also advisable to pass both the input and output data through the same low-pass or band-pass filter that has a passband covering the interesting frequencies before starting the estimation process. The effects of low-frequency disturbances, drift, and so on, are thereby reduced. So the focus on the model fit is also automatically moved to the most important frequency bands. Sometimes linear models that are estimated from data are normally based on signals measured relative to an equilibrium point. If we work with a tailor-made model with absolute signal levels built in, the signal should of course be kept at their physical units. A rule of thumb is to subtract at least the mean level in each measured signal. The mean levels also often drift away during the experiment. This can, be eliminated by high-pass filtering. High-frequency noise in measurements in uninteresting frequency bands reflects improper selection of the sampling interval and the anti-aliasing filter. This kind of noise may be removed by low-pass filtering, possibly followed by decimation. [7]

5.2 Model Structure Selection [5, 7, 8]

Model structure selection for a given physical system is perhaps the most difficult decision to make in system identification. In general, tailor-made model is based on known physical relationships and is parsimonious with parameters. A direct physical interpretation can always help us decide whether the estimates are reasonable. So, this type model can be high quality, that is, small bias and variance inaccuracy. The disadvantages for this modeling procedure are time consuming and computationally demanding in the minimization process. Also, it may contain many unknown parameters whose individual values are not essential for the process. The decision of choosing a tailor-made model depends on whether the identification software package supporting for such an exercise.

So we may always try the simple model first, i.e., ready-made models. Then the more complex ones only if the simple models fail. Which one should we choose? There is no general answer, but thinking through the system's function and basic physics will help. In practice we can test and compare different structures for their characteristics. Some general points of view are listed below.

    (1) The ARX model is the easiest one to use. The foremost disadvantage is that the disturbance model $H(z^{-1}, \theta) = 1/A(z^{-1})$ comes along with the system's poles. This may cause an incorrect estimate because the $A(z^{-1})$ polynomial also has to describe the disturbance properties. Higher orders in $A(z^{-1})$ and $B(z^{-1})$ than necessary may be required. If the signal-to-noise ratio is high, then the above mentioned disadvantage will be less important.

    (2) The ARMAX model is the most often used model and gives us extra flexibility to handle disturbance modeling.

    (3) Using OE model we can describe the system dynamics separately from the noise. If the system operates without feedback during the data collecting, a correct description of $B(z^{-1})/F(z^{-1})$ can be achieved. However, minimization process will be more involved than that for ARMAX model.

    (4) The BJ model is the "complete model" in which the disturbance properties are modeled separately from the system dynamics.

Since models ARX and ARMAX have the same dynamics, they are suitable if the dominating disturbances enter at the input. The BJ model is preferable if modeling disturbances come in as measurement noise in the output.

Another issue is how to compare models in different structures. One suggested way is to evaluate prediction error variances of the models with new data sequences. In statistical terms this is called cross validation. If a model is able to predict a fresh set of data better than another, it should be treated as a better model. The obvious disadvantage with the method is a set of fresh data has to be reserved for model comparison, so less available information can be used for estimation.

If the prediction error comparison has to be made on the basis of the data that already have been used in the model estimation, a few problems emanate. At first, a larger model will always give a lower value of the criterion function, since it has been obtained by minimizing over more parameters. Secondly, "overfit" phenomenon has been observed. This happens because we use unnecessary parameters to fit the specific disturbance signals in the data set. Unfortunately, this extra effort will not serve any purpose because the use of the model will always be under different disturbances. Several methods have been proposed to find a balance between "model fit" and "number of parameters involved" in the literature.[32,33] They are

(1) Akaike's information criterion (AIC):

$$\min_{d,\theta}(1+\frac{2d}{N})\sum_{k=1}^{N}\varepsilon^2(k,\theta)$$

(2) Final prediction error (FPE):

$$\min_{d,\theta}(\frac{1+d/N}{1-d/N})\frac{1}{N}\sum_{k=1}^{N}\varepsilon^2(k,\theta)$$

(3) Rissanen's minimal description length

$$\min_{d,\theta}(1+\frac{2d}{N}\log N)\sum_{k=1}^{N}\varepsilon^2(k,\theta)$$

Here N is the number of data, and d is θ's dimension (the number of estimated parameters).

Model comparison can be carried out through digital simulation.[34] For linear models it is also instructive to compare their Bode diagrams and pole-zero diagrams.[15] A comparison of spectrum between measured data and those obtained from different models is also very useful.[35]

5.3 Order & Delay Determination

The order of a selected model is not a priori information. However, it is possible to estimate the order of the model from the samples of the impulse response by calculating the determinant of the Hankel matrix. [36] Woodside in 1971 proposed a method which used the rank of the product moment matrix to calculate the rank of a linear model.[37, 38]

Ljung presents a procedure to determine the order and delays of a ready-made model.[7] Since it is quite useful, we repeat it as follows;

1) First get a reasonable estimate of the delay by correlation analysis and/or by testing all reasonable ones in a, say, fourth-order ARX model. Pick the delay that gives the best model performance (sum of squared prediction errors on a validation set).

2) Then test many ARX models of different orders with this delay. Pick the order that give the best model performance.

3) This model may be of unnecessarily high order to describe the system dynamics, since the poles of an ARX model also describe the noise properties. Thus plot the zeros and the poles of the resulting model (with uncertainty regions marked) and look for cancellations. The surviving numbers of poles and zeros give us indications of the necessary order for the dynamics from input to output. Then try ARMAX, OK, or BJ-models with this order for G(z-1) =B(z-1)/F(z-1) and first- or second-order models for the noise characteristics.

5.4 Model Validation

Model validation is a process to investigate whether the model identified can be accepted for the intended use. After we have spent efforts in finding a model of least possible complexity within the limits of required accuracy, this process will provide us with confidence of the model quality. We can do this by comparing the model's output to the measured one on data sets that were not used for the fit. If approximately the same properties are obtained under such varied conditions, we should feel confident that the model has some significant features of the system.

Residual Analysis have been developed and used for testing the suitability of a model. A model is considered satisfactory if the residuals form a white-noise sequence with zero mean, and as small a variance possible If the sequence of residuals satisfies the test for whiteness, then we select the model for which this sequence has the minimum variance. [33, 38]

6 CONCLUSIONS

System identification has been proved as a convenient and useful tool for model building in many different fields. More than 10,000 successful applications have been published in the literature. These cover widely varying areas from process industry, ship dynamics, signal processing, and seismology to biomedicine, ecology, and econometrics. In this paper we have briefly reviewed the basic concepts of system identification, the available algorithms, and possible problems and suggestions that we may encounter in the experiment design. We hope this work will help readers make sagacious judgement in the selection and use of software packages and identify new applications in the field of civil engineering.

REFERENCE

[1] H.H. Kagiwada, System Identification Methods and Applications, Addison-Wesley, Reading, Massachusetts, 1974

[2] L. Ljung, The System Identification Toolbox, User's Guide, 3rd ed. The MathWorks,Inc., Natick, Mass., 1992.

[3] K.J. Astrom, P. Eykhoff, "System Identification- A Survey," Automatica, Vol.7,1971, pp.123-162

[4] R. Isermann, edt., System Identification, IFAC Tutorials, Pergamon Press, 1981 ???? [5] R. Johansson, System Modeling Identification, Prentice-Hall, Englewood Cliffs, NJ, 1993

[6] J. Juang, Applied System Identification, Prentice-Hall, Englewood Cliffs, NJ, 1994 [7] L. Ljung, Modeling of Dynamic Sytems, Prentice-Hall, Englewood Cliffs, NJ, 1994

[8] N.K. Sinha, B. Kuszta, Modeling and Identification of Dynamic Systems, Van Nostrand Reinhold Company, New York, 1983

[9] T.C. Hsia, System Identification, Lexington Books, Toronto, 1977 [10] A.P. Sage, J.L. Melsa, System Identification, Academic Press, New York, 1977

[11] W.L. Brogan, Modern Control Theory, Prentice-Hall, Englewood Cliffs, NJ, 1985 [12] C.M. Close, D.K. Frederick, Modeling and Analyses of Dynamic Systems, Houghton Mifflin, Boston, 1993

[13] J.C. Santamarina, D. Fratta, Introduction to Discrete Signals and Inverse Problems in Civil Engineering, ASCE Press, Reston, VA, 1998

[14] C.L. Phillips, H.T. Nagle, Digital Control System Analysis and Design, 3rd edition, Prentice-Hall, Englewood Cliffs, NJ, 1995

[15] B.C. Kuo, Automatic Control Systems, 7th edition, Prentice-Hall, Englewood Cliffs, NJ, 1995 [16] J.E. Slotine, W. Li, Applied Nonlinear Control, Prentice Hall, Englewood Cliffs, N.J., 1995

[17] H. Kagiwada, R. Kalaba, B. Vereeke, "Invariant Imbedding and Fredholm Integral Equations with Displacement Kernels on an Infinite Interval," Int. J. Comput. Math., Vol.2, 1970, pp.221-229 [18] H. Kagiwada, R. Kalaba, "An Initial Value Method for Fredholm Resolvents of Seidegenerate Kernals," J. Optimiz. Th. Appl., Vol.11, No.5, 1973, pp.517-532

[19] W. Ishak, N.K. Sinha, "Time Domain Approximation and System Modelling Using a Damped Least-squares Algorithm," Int. J. of Systems Science, Vol.7, 1976, pp.635-640

[20] N.K. Sinha, A. Sen, R. To, "Time Domain Approximation Using Digital Methods," Int. J. of Systems Science, Vol.5, 1974, pp.373-382

[21] K. Ogata, Discete-Time Control Systems, Prentice-Hall, Englewood Cliffs, NJ, 1987 [22] D. Graupe, Time Series Analysis, Identification and Adaptive Filtering, Krieger, Malabar, FL., 1984

[23] R.G. Brown, Introduction to Random Signal Analysis and Kalman Filtering, John Wiely & Sons, New York, 1983

[24] B. Widrow, S.D. Stearns, Adaptive Signal Processing, Prentice-Hall, Englewood Cliffs, N.J., 1985 [25] A. Cichocki, R. Unbehauen, Neural Networks for Optimization and Signal Processing, John Wiely & Sons, New York, 1993

[26] K.S. Narendra, K. Parthasarathy, "Identification and Control of Dynamical Systems Using Neural Networks," IEEE Trans. Neural Networks, Vol.1, 1990, pp.4-27 [27] J.S. Bendat, Nonlinear System Analysis & Identification from Random Data, Wiley, New York, 1990

[28] R.K. Mehra, "Frequency Domain Synthesis of Optimal Inputs for Linear System Parameter Estimation," Trans. of ASME, JDMC, Vol.98, Ser. G, No.2, 1976, pp, 130-138

[29] R.K. Mehra, "Optimal Input Signal for Parameter Estimation in Dynamic System- Survey and New Results," IEEE Trans. Automatic Control, Vol.98, 1974, pp.753-768

[30] G.C. Goodwin, R.L. Payne, Dynamic System Identification: Experiment Design & Data Analysis, Academic Press, New York, 1977

[31] R.M. Gagliardi, "Input Selection for Parameter Identification in Discrete Systems," IEEE Trans. Automatic Control, Vol.12, 1967, pp.597-599

[32] J. Rissanen, "Modeling by Shortest Data Description," Automatica, Vol.14, 1978, pp.465-471

[33] H. Akaike, " A New Look at the Statistical Model Identification," TAC-19, 1977, pp.718-723

[34] J.S. Rosko, Digital Simulation of Physical Systems, Addison-Wesly, Reading, Mass., 1972

[35] J.S. Bendat, A.G. Piersol, Engineering Applications of Correlation and Spectral Analysis, Wiley, New York, 1980

[36] R.C.K. Lee, Optimal Estimation, Identification, and Control, M.I.T. Press, Cambridge, Mass. 1964

[37] C.M. Woodside, "Estimation of the order of linear system," Automatica, Vol.7, 1971, pp.727-733

[38] G.E.P. Box, G.M. Jenkins, Time Series Analysis, Forecasting and Control, 2nd Ed. Holden-Day, San Francisco, 1989.

[39] P. Stoica, "A Test for Whiteness," IEEE Transactions on Automatic Control, Vol.22, 1977, pp.992-993

Artificial Intelligence and Mathematical Methods in Pavement and Geomechanical Systems, Attoh-Okine (ed.)
© *1998 Balkema, Rotterdam, ISBN 90 5809 028 0*

Author index